The Brain

Edited by Kenneth Partridge

The Reference Shelf
Volume 81 • Number 1
The H.W. Wilson Company
New York • Dublin
2009

Preface

The human brain has been called the most complex structure in the known universe. Weighing just three pounds, it comprises some 100 billion nerve cells, or neurons, which together form a vast network of connections. The brain is often likened to a computer, but in fact it is far more powerful and versatile than any machine yet designed by man. In addition to processing sensory stimuli, controlling movement, and regulating unconscious bodily functions, such as heartbeat, the brain is the headquarters of our intellect and creativity. It has enabled man to invent the wheel, build the pyramids, paint the *Mona Lisa*, write *Hamlet*, explore the moon, and devise the theory of relativity. These and other astounding achievements have had much to do with the brain's capacity for language, which allows us to communicate with each other and make sense of natural phenomena in ways other animals cannot. The brain is also in charge of emotions and memory, and even though people speak of love, honor, and sincerity as provinces of the heart, they, like all feelings, are products of the brain.

Throughout much of human history, science yielded few insights regarding the inner workings of the brain. While researchers spent centuries drawing diagrams and dreaming up elaborate names for specific anatomical regions, it wasn't until the 20th century that neuroimaging technology began to answer questions regarding the brain's mechanics. Thanks to breakthroughs such as the advent of fMRI, or functional magnetic resonance imaging, researchers are now capable of monitoring the brain activity of living subjects. As a result we now know a great deal about where the brain processes various types of information. For as far as science has come, however, many question marks remain. Chief among these is consciousness. Though thinkers have offered countless theories, no one has yet been able to explain what gives us our sense of self, the feeling that we're freethinking beings in control of our bodies. If the 1990s were the "Decade of the Brain," as President George H. W. Bush once declared, the 2010s could well become the "Decade of the Mind," a period in which scientists look beyond the brain's functionality and tackle this most fundamental—and confounding—of human questions.

This volume of the Reference Shelf series examines what modern science has taught us about the brain and considers what remains to be discovered. The first chapter, "Structure, Functions, and Historic Approaches to Study," provides both a brief history of neuroscience and a summary of the brain's anatomy. The sec-

ond chapter, "Mapping the Minds: Advances in Brain-Imaging Technology," analyzes how computer technology has helped researchers match brain structures to
specific functions.

In the third chapter, "Works in Progress: Brain Development from Infancy to
Adulthood," selected articles focus on the brain's "plasticity," or ability to change
with time. This idea is relatively new, as for years scientists believed that brains
stop developing and adding new neurons after infancy. The fourth chapter, "How
the Brain Processes Language," focuses on what is perhaps the most essential of
all human traits—the capacity for verbal and written communication.

Articles in the fifth chapter, "The Brain and Aging," examine how the brain
changes over the years. While diseases such as Alzheimer's can lead to memory loss
and dementia, emerging evidence suggests that, with respect to certain types of
problem solving, the brain can actually improve with age. The final chapter, "The
'Hard Problem': Efforts to Understand Human Consciousness," presents various
theories on a question that continues to baffle scientists. While some thinkers have
suggested that consciousness is an illusion, and that our minds essentially run on
autopilot, others refuse to discount the roles that "we" play in our own lives.

In conclusion, I would like to thank the many authors and publishers who have
granted permission to reprint the articles contained in this book. In addition, I
extend my gratitude to my many friends and colleagues at the H.W. Wilson Company, especially Paul McCaffrey and Richard Stein.

Kenneth Partridge
February 2009

1

Structure, Functions, and Historic
Approaches to Study

Editor's Introduction

As far as historians can tell, human study of the brain dates back to the 27th century B.C., when the ancient Egyptian engineer and physician Imhotep started poking around inside the skulls of injured patients. Then—as is still the case today—the brain was a mysterious organ, and some 2,000 years later, even the great thinker Aristotle struggled to explain its function. He believed the heart to be the control center of the body, a notion that wasn't disproved until the 2nd century A.D., when another Greek scientist, Galen, finally recognized the importance of the brain. While Galen's discovery—the result of grisly experiments on live pigs—constituted a major breakthrough, it spawned far more questions than it answered. For the next 17 centuries, the best scientists could do was catalog the brain's anatomy, drawing increasingly detailed diagrams of structures whose functions they were woefully ill-equipped to study.

Thanks to 20th century technological advances, such as the advent of fMRI (functional magnetic resonance imaging), a technique that makes it possible to observe the brain activity of living subjects, scientists now know a great deal more about how the brain works. With increased knowledge has come renewed interest in the brain's anatomy, as researchers finally have the tools necessary to pinpoint the specific regions responsible for our thoughts, actions, senses, and even emotions.

The articles in this chapter provide an overview of the brain's basic structures, as well as a brief history of man's efforts to understand the inner workings of his or her own head. In the first article, "The Strange Anatomy of the Brain," David Bainbridge charts the developments that led from Imhotep's curious fumbling to the relative precision of today's fMRI scans. He posits that the human brain provides strong evidence for evolution, since it shares much in common with those found in other primates. Bainbridge suggests that the sheer size of the human brain, at least in proportion to our bodies, might explain why we've developed language and other mental abilities that other creatures have not.

The second piece, "Brainbox," explains how a network of 100 billion nerve cells transmits information inside the brain. Again making reference to evolution, the article examines how the human brain differs from those found in reptiles and rats. What sets humans apart, the article concludes, is the cerebral cortex, which accounts for 80 percent of our brain's mass. That region is responsible for such crucial functions as vision, movement, and language.

While many scientists marvel at the complexity and computational power of the human brain, others aren't as reverent. "In Our Messy, Reptilian Brains," the final article in this chapter, Sharon Begley considers why the brain might actually be a "cobbled-together mess," to quote David Linden, a neuroscience professor at Johns Hopkins University. Begley likens the human brain to "an iPod built around an eight-track cassette player," since, in addition to highly advanced, uniquely human structures, it contains many antiquated features found in the brains of our reptilian ancestors. In keeping with Linden's assertion that the brain is "quirky, inefficient and bizarre," Begley ponders the narrative nature of dreams, something science has yet to explain.

The Strange Anatomy of the Brain[*]

By David Bainbridge
New Scientist, January 26, 2008

Human curiosity about the workings of the brain dates back at least 46 centuries. The first appearance of the word "brain" is on the Edwin Smith Surgical Papyrus, an Egyptian manuscript dating from around 1600 BC but thought to have been copied from the writings of Imhotep, an engineer, architect and physician who lived 1000 years earlier. The papyrus is the earliest known work on trauma medicine, and among other things it describes head injuries.

We do not know whether those injuries were suffered during the carnage of war or in the chaos of an ancient building site, but they make sobering reading: men who are paralysed, men who can only crouch and mumble, and men whose skulls are split open to reveal the "skull offal" inside, convoluted like "the corrugations . . . in molten copper". The author marvels at the opportunity to study this most mysterious of organs, and describes how his patients start to shudder when he thrusts his fingers into their wounds.

Today we know so much about the brain it is easy to forget that for much of human history its workings were entirely hidden from view. For centuries it was the preserve of anatomists who catalogued and mapped its internal structure in exquisite detail even though they had little idea what any of the structures actually did. Only now, as we finally gain access to the brain's inner workings, have those brain maps from the past started to make real sense.

Two millennia after Imhotep, the ancient Greeks were less enamoured of skull offal. Aristotle placed the control centre of the body at the heart, not the brain, presumably because it is demonstrably physically active, diligently pulsing throughout life. He found the brain to be still, and erroneously described it as "bloodless, devoid of veins, and naturally cold to the touch". This coldness, as well as its corrugated surface, led Aristotle to suggest that the brain was merely a radiator, dissipating the heat generated by the heart.

It was Galen, a second-century Greek physician, who finally established the crucial role of the brain in controlling the body. Much of his evidence came from detailed anatomical study of animal brains. Galen was impressed by the brain's complexity—its anatomy is far more complicated than that of any other organ—which suggested to him that it must be doing something important. He thought its purpose was to interact with the sense organs in the head.

Galen was a practical man, and he confirmed his ideas about brain function with public experiments on live animals. He describes with great relish how, having tracked the nerves from a pig's brain to its voice box, he could expose them in the neck of a living, squealing animal, only to silence it with a cut. These ghoulish experiments showed that the brain controls the body, and Galen went on to argue that the brain and its nerves are also responsible for sensation, perception, emotion, planning and action.

Galen's work on brain function remained the state of the art right up to the 20th century. In the absence of new experimental techniques there was little more that could be done to study its inner workings. Yet the brain remained of great interest to anatomists, and over the course of the intervening 17 centuries they methodically mapped and catalogued its structure in ever finer detail. The impressive complexity of all those parts must have screamed out the idea that the brain was doing something very complicated, but what exactly? There was no way to find out.

This mapping and cataloguing of the brain's many structures has left a legacy of wonderfully descriptive and evocative names—almonds, sea horses, hillocks, girdles, breasts and "black stuff"—all the more so for being expressed in Latin and Greek. There are mysterious regions bearing the names of their otherwise-forgotten discoverers, such as the tract of Goll, the fields of Forel, Monro's holes and the radiations of Zuckerkandl. Others seem inexplicable (brain sand), starkly functional (the bridge) or ludicrously florid (nucleus motorius dissipatus formationis reticularis, which translates as "the dispersed motor nucleus of the net-like formation"). Some are just plain defeatist (substantia innominata or "unnamed stuff").

Fanciful as many of these names are, once they had been chosen they tended to stick, and many have survived into the exacting world of modern neuroscience. Memory researchers now probe the molecular machinery of the sea horse (hippocampus); scientists studying emotion scan the almonds (amygdalae) for flickers of fear. The modern geography of the brain has a deliciously antiquated feel to it—rather like a medieval map with the known world encircled by terra incognita where monsters roam.

ANTIQUATED FEEL

The old names bear little relation to our understanding of brain function, so navigating around the brain can be an arcane pursuit. Yet the quaintly outdated nomenclature has scarcely held science back, and in some respects this detailed mapping of the brain has helped clinicians, evolutionary biologists, philosophers and others make sense of it. The structure of the brain has always formed the core of what we know about the mind—after all, its anatomy remains the only thing about it that we do know for sure.

Take evolutionary biology: the structure of the brain is excellent evidence for evolution, and of our unexceptional place in the scheme of things. Before Darwin, it must have seemed inexplicable that humans, the divinely chosen overseers of the Earth, should have brains almost indistinguishable from those of dumb beasts. No one had been able to find an obvious structural difference in the human brain which explained our intelligence, language, wisdom and culture. But that did not stop anatomists searching for it. In 1858, a year before *On the Origin of Species*, a small protrusion into one of the brain's inner cavities—called either the hippocampus minor (little sea horse) or calcar avis (cockerel's spur)—was hailed as the definitive distinguishing feature of the human brain. Yet almost immediately supporters of the theory of evolution demonstrated that this little structure is also present in many other primates, and the uniqueness of the human brain vanished once more.

As it turns out, all parts of the human brain are also present in other primates, and we share the general plan of our brain with all backboned animals. A trout's brain is made up of the same three major regions as a human brain, and many of its parts have similar functions to their human equivalents. These structural similarities are reassuring to evolutionary biologists because they imply that the human brain evolved by the same processes that generated all other animals' brains. Yet confusingly, they leave us with no obvious location for the abilities we think of as distinctively human.

Vertebrates probably share a common brain architecture for two reasons. First, once an ancient ancestral vertebrate had a brain, it seems unlikely that evolution would have any reason to dismantle it and start afresh. Little surprise, therefore, that human brains conform to a standard vertebrate plan. The other reason is that, despite our different environments and ways of life, all animals have to process the same types of information. No matter how bizarre a vertebrate is, it receives only three types of incoming sensory data: chemical (smell and taste), electromagnetic (light, and electric and magnetic fields) and movement (touch and sound). This restricted sensory palette may have been what gave rise to the three-part vertebrate brain, with the front part processing smell information, the middle dealing with vision and the back interpreting sound. The laws of physics have never changed, so no new senses ever appeared and no new segments were added to our brains.

This still leaves us with the problem of how humans can be so intelligent when their brain has the same arrangement as a trout's. At this point all those centuries of anatomical cataloguing come in handy, because they tell us how heavy different animals' brains are. We like to think that human brains, at 1300 grams, are unusually large; indeed, comparison with the 500-gram chimp brain bears this out. However, bottlenose dolphin brains weigh in at around 1600 grams, while sperm whales are cerebral giants at 8000 grams, so sheer size is not the only thing that matters. However, in most mammals there is a strong mathematical relationship between body size and brain size: the weight of an animal's brain can usually be predicted with a high degree of accuracy from the weight of its body. This is why whales, for example, have such big brains. But not so with humans: our brains are at least three times as large as would be expected from the calculations, suggesting that we have evolved tremendous cerebral overcapacity.

This huge expansion of the human brain, most of which occurred within the last 10 million years, suggests that the differences between the mental abilities of humans and those of other animals are down to brain size: quantity, not quality. Perhaps we have passed a "critical mass" at which language, abstraction and all our other cognitive abilities can start to develop.

An additional trend in human brain evolution has been the way in which many functions have been shunted to novel locations. For example, sensory processing and control of movement have been squashed ever higher into the upper regions of our brain—the cerebral cortex. The huge human cortex may look like a larger version of any other animal's, but it is stuffed full of so many different functions that it may have come to work in a fundamentally different way.

ANATOMY RETURNS

Somewhat ironically, the one discipline where brain anatomy has not always held centre stage is brain science itself. Once the anatomists had completed their work in the early 20th century, neuroscientists promptly turned their backs on it as function superseded structure at the cutting edge of research. In the past few years, however, we have come to the realisation that to understand the function of the brain we must also understand its complex structure. Brain anatomy is in the ascendant again, and here are two reasons why.

The first is a practical, clinical one. Medical and veterinary students of my generation were often perplexed by the amount of time we spent studying brain anatomy, especially stained cross sections cut from the brains of dead people and animals. After all, the brain is invisible on radiographs and for most clinical purposes it was treated as a black box hiding inside the impenetrable skull. What use could all that anatomy possibly have?

However, an invention that has come along since I finished my clinical training has changed all that. Magnetic resonance imaging (MRI) scanners detect the radio pulses released by spinning protons as they realign in a magnetic field after they are knocked about by a pulse of radio waves. This phenomenon can be used to map the anatomy of the living brain, giving access to its workings for the first time. The information yielded by MRI comes in the form of computer-generated cross sections through the brain: suddenly, the old brain slices have taken on a new importance as part of the everyday process of diagnosing and treating disease.

The other reason why brain anatomy is important once again is a more philosophical one. Looking back over a century of studying brain function, one lesson is clear: we cannot truly understand a mental process until we know where it occurs. We now understand many complex brain processes—something that would have seemed incredible a few decades ago—and in every case that understanding followed our discovery of where the activity takes place. Memories are manipulated in the sea horse, fear in the almonds, vision is processed near the cockerel's spur, and hearing in the hillocks (colliculi). Once neuroscientists know where a process occurs, they can start to pick it apart and find out how it occurs.

The importance of place is also illustrated by a counter-example—consciousness. We are pretty sure that consciousness takes place in the brain and that it must have evolved by the same mechanisms as every other cerebral process. Beyond that we are stumped. There are many theories of the nature of consciousness, but they are often difficult to test. A major reason for this is that we do not know where consciousness happens, so we cannot study the brain regions which generate it. Of course, there may be no single area of the brain that produces consciousness—it may be widely distributed across many regions—but again we do not know what those regions are. Until we do, consciousness will remain stubbornly resistant to our investigations.

Brain anatomy is in vogue again. For much of the time humans have pondered the nature of the brain, its anatomy was all we knew. In the 20th century it looked as if function had overtaken structure as the best way to understand the brain, but we can now see that the two are inextricably linked. More than anywhere else in the body, in the brain a sense of geography is crucial. There are many fantastic journeys ahead.

Brainbox[*]

A History and Geography of the Brain

The Economist, December 23, 2006

The reason that people have brains is that they are worms. This is not a value judgment but a biological observation. Some animals, such as jellyfish and sea urchins, are radially symmetrical. Others are bilaterally symmetrical, which means they are long, thin and have heads.

Headless animals have no need for brains. But in those with a head the nerve cells responsible for it—and thus for sensing and feeding—tend to boss the others around. That still happens even when a long, thin, animal evolves limbs and a skeleton. Bilateralism equals braininess.

A healthy human brain contains about 100 billion nerve cells. What makes nerve cells special is that they have long filamentary projections called axons and dendrites which carry information around in the form of electrical pulses. Dendrites carry signals into the cell. Axons carry signals to other cells. The junction between an axon and a dendrite is called a synapse.

Information is carried across synapses not by electrical pulses but by chemical messengers called neurotransmitters. One way of classifying nerve cells is by the neurotransmitters they employ. Workaday nerve cells use molecules called glutamic acid and gamma aminobutyric acid. More specialised cells use dopamine, serotonin, acetylcholine and a variety of other molecules. Dopamine cells, for example, are involved in the brain's reward systems, generating feelings of pleasure.

Many brain drugs, both therapeutic and recreational, work either by mimicking neurotransmitters or altering their activity. Heroin mimics a group of molecules called endogenous opioids. Nicotine mimics acetylcholine. Prozac promotes the activity of serotonin. And cocaine boosts the effect of dopamine, which is one reason why it is so addictive.

Apart from specialised nerve cells, there is a lot of anatomical specialisation in the brain itself. Three large structures stand out: the cerebrum, the cerebellum and the brain stem. In addition, there is a cluster of smaller structures in the middle. These are loosely grouped into the limbic system and the basal ganglia, although not everyone agrees what is what.

Most brain structures, reflecting the bilateral nature of brainy organisms, are paired. In particular, the cerebrum is divided into hemispheres whose only direct connection is through three bundles of nerves, the most important of which is called the corpus callosum. (Many parts of the brain have obscure Latin names.)

This anatomical division of the brain reflects its evolutionary history. The brains of reptiles correspond more or less to the structures known in mammals as the brain stem and cerebellum. In mammals the brain stem is specialised for keeping the hearts and lungs working. The cerebellum is for movement, posture and learning processes associated with these two things. It is the limbic system, basal ganglia and cerebrum that do the interesting stuff that distinguishes mammalian brains from those of their reptilian ancestors.

The limbic system is itself divided. Some of the main parts are the hippocampus, the amygdala, the thalamus and the hypothalamus. The largest of the basal ganglia is the caudate. The pineal gland, which lies behind the limbic system, is the only brain structure that does not come in pairs. The 17th-century French philosopher René Descartes thought it was the seat of the human soul.

Descartes, however, was wrong. It is in fact the cerebrum's outer layer, the cerebral cortex that is man's true distinguishing feature. The cerebral cortex forms 80% of the mass of a human brain, compared with 30% of a rat's. It is divided into lobes, four on each side. The rearmost one, called the occipital, handles vision. Then come the parietal and temporal lobes, which deal with the other senses and with movement. At the front, as you would expect, is the frontal lobe.

This is humanity's "killer app", containing many of the cognitve functions associated with human-ness (although that most characteristic human function, language, is located in the temporal and parietal lobes, and only on one side, usually the left). Man's huge frontal lobes are the reason for the species' peculiarly shaped head. No wonder that in English-speaking countries the brainiest of the species are known as "highbrow".

In Our Messy, Reptilian Brains[*]

By Sharon Begley
Newsweek, April 9, 2007

Let others rhapsodize about the elegant design and astounding complexity of the human brain—the most complicated, most sophisticated entity in the known universe, as they say. David Linden, a professor of neuroscience at Johns Hopkins University, doesn't see it that way. To him, the brain is a "cobbled-together mess." Impressive in function, sure. But in its design the brain is "quirky, inefficient and bizarre . . . a weird agglomeration of ad hoc solutions that have accumulated throughout millions of years of evolutionary history," he argues in his new book, "The Accidental Mind," from Harvard University Press. More than another salvo in the battle over whether biological structures are the products of supernatural design or biological evolution (though Linden has no doubt it's the latter), research on our brain's primitive foundation is cracking such puzzles as why we cannot tickle ourselves, why we are driven to spin narratives even in our dreams and why reptilian traits persist in our gray matter.

Just as the mouse brain is a lizard brain "with some extra stuff thrown on top," Linden writes, the human brain is essentially a mouse brain with extra toppings. That's how we wound up with two vision systems. In amphibians, signals from the eye are processed in a region called the mid-brain, which, for instance, guides a frog's tongue to insects in midair and enables us to duck as an errant fastball bears down on us. Our kludgy brain retains this primitive visual structure even though most signals from the eye are processed in the visual cortex, a newer addition. If the latter is damaged, patients typically say they cannot see a thing. Yet if asked to reach for an object, many of them can grab it on the first try. And if asked to judge the emotional expression on a face, they get it right more often than chance would predict—especially if that expression is anger.

They're not lying about being unable to see. In such "blindsight," people who have lost what most of us think of as vision are seeing with the amphibian visual system. But because the midbrain is not connected to higher cognitive regions, they have no conscious awareness of an object's location or a face's expression. Consciously, the world looks inky black. But unconsciously, signals from the midbrain are merrily zipping along to the amygdala (which assesses emotion) and the motor cortex (which makes the arm reach out).

Primitive brains control movement with the cerebellum. Tucked in the back of the brain, this structure also predicts what a movement will feel like, and sends inhibitory signals to the somatosensory cortex, which processes the sense of touch, telling it not to pay attention to expected sensations (such as the feeling of clothes against your skin or the earth beneath your soles). This is why you can't tickle yourself—the reptilian cerebellum has kept the sensation from registering in the feeling part of the brain. Failing to register feelings caused by your own movements claims another victim: your sense of how hard you are hitting someone. Hence, "but Mom, he hit me harder!"

Neurons have hardly changed from those of prehistoric jellyfish. "Slow, leaky, unreliable," as Linden calls them, they tend to drop the ball: at connections between neurons, signals have a 70 percent chance of sputtering out. To make sure enough signals do get through, the brain needs to be massively interconnected, its 100 billion neurons forming an estimated 500 trillion synapses. This interconnectedness is far too great for our paltry 23,000 or so genes to specify. The developing brain therefore finishes its wiring out in the world (if they didn't, a baby's head wouldn't fit through the birth canal). Sensory feedback and experiences choreograph the dance of neurons during our long childhood, which is just another name for the period when the brain matures.

With modern parts atop old ones, the brain is like an iPod built around an eight-track cassette player. One reptilian legacy is that as our eyes sweep across the field of view, they make tiny jumps. At the points between where the eyes alight, what reaches the brain is blurry, so the visual cortex sees the neural equivalent of jump cuts. The brain nevertheless creates a coherent perception out of them, filling in the gaps of the jerky feed. What you see is continuous, smooth. But as often happens with kludges, the old components make their presence felt in newer systems, in this case taking a system that worked well in vision and enlisting it [in] higher-order cognition. Determined to construct a seamless story from jumpy input, for instance, patients with amnesia will, when asked what they did yesterday, construct a story out of memory scraps.

It isn't only amnesiacs whose brains confabulate. There is no good reason why dreams, which consolidate memories, should take a narrative form. If they're filing away memories, we should just experience memory fragments as each is processed. The cortex's narrative drive, however, doesn't turn off during sleep. Like an iPod turning on that cassette player, the fill-in-the-gaps that works so well for jumpy eye movements takes the raw material of memory and weaves it into a coherent, if bizarre, story. The reptilian brain lives on.

2

Mapping the Mind: Advances in Brain-Imaging Technology

Editor's Introduction

In July 1990 President George H. W. Bush designated the 1990s the "Decade of the Brain." His proclamation was based on recent advances in neuroscience and a belief that still more breakthroughs were on the horizon. Indeed, much of what scientists now know about the human brain is the result of studies conducted over the last two decades. Thanks to cutting-edge computer-imaging techniques, such as fMRI (functional magnetic resonance imaging) and PET (positron emission tomography) and CAT (computed axial tomography) scans, scientists now have the ability to monitor the brains of living subjects and record, in "real time," how different regions respond to stimuli. Such "neuroimaging" tests have helped researchers identify which parts of the brain are responsible for performing various tasks—such as recognizing faces—and determining complex emotions and character traits. In the words of neurology professor Richard Restak, we've entered the era of the "New Brain," a golden age of experimentation that has brought about "a revolutionary change in our understanding."

The selections in this chapter discuss how neuroimaging has revolutionized the study of the brain and provided new insights into what makes us think, act, and feel the way we do. In the first piece, "The New Brain," Restak explores how technology has transformed "cognitive science," which he defines as "the study of the brain mechanisms responsible for our thoughts, mood, decisions, and actions." Restak focuses on how brain research has yielded "practical applications that can be put to use in our daily lives." In the future, he predicts, studies will lead to better treatments for mental illness and a richer understanding of how violence and other forms of stress negatively impact the brain.

In her two-part series "Unlocking the Secrets of the Brain," Tabitha M. Powledge provides a detailed overview of today's leading neuroimaging techniques, including fMRI, PET, and SPECT (single photon emission computed tomography). She also considers the advantages of EEG (electroencephalogram) testing, an older technology that measures brain waves. Powledge opens her series with the oft-told story of Phineas Gage, a Vermont railroad worker, who in 1848 was struck in the head by a 13-pound piece of iron. The rod pierced the frontal lobe of his brain, and though he survived the accident, he experienced a drastic change in personality, giving scientists one of their first glimpses into how the brain controls behavior.

In "MRI: A Window on the Brain," Paul Raeburn considers how advances in MRI testing might one day help doctors diagnose and treat complex psychiatric ailments, such as bipolar disorder. Such conditions are notoriously difficult to diagnose, since the brains of those afflicted tend to look similar to those of normal patients. Thanks to a new technique known as MRI spectroscopy, Mayo Clinic researcher John Port is honing in on a diagnostic test for bipolar disorder, one that, if successful, could be applicable to other diseases. Columbia University psychiatrist Bradley Peterson, meanwhile, is using MRI to study children born prematurely. His findings suggest that premature children begin life with abnormally large ventricles, or brain cavities, and go on to have smaller brains.

Scientists aren't the only ones interested in neuroimaging technology. Economists have long studied how consumers make decisions, and in his article "The Economics of Brains," the final entry in this chapter, Gregory T. Huang explores the up-and-coming field of "neuroeconomics." While traditional economic theory maintains that people act rationally, neuroeconomics seeks to explain why someone might engage in risky behavior, such as gambling away a paycheck. The article ends with a discussion of "neuromarketing." In the future, Huang writes, companies may use neuroimaging research to target pleasure sensors in our brains and craft more effective advertisements.

The New Brain[*]

By Richard Restak
The Futurist, January/February 2004

We have learned so much about the human brain during the past two decades that it's fair to speak of a revolutionary change in our understanding. The era of the Old Brain is giving way to that of the New Brain.

The Old Brain was remote and mysterious, deeply hidden within the skull and inaccessible except to specialists daring enough to pierce its three protective layers. Thanks to that inaccessibility and the risks involved in plumbing its depths, brain experts knew little about the functioning of the normal brain; they certainly searched in vain for answers to such fascinating questions as, "How is the brain related to our everyday thoughts, emotions, and behavior?"

The New Brain, in contrast, does not require dangerous intrusions but can now be depicted using sophisticated computer-driven imaging techniques such as CAT and PET scans and MRIs. These techniques reveal exquisitely subtle operational details and provide windows through which neuroscientists can view different aspects of brain functioning without opening the skull or performing other risky procedures.

Thanks to the development of new imaging technologies, brain science is capable of providing us with insights into the human mind that only a few decades ago would have been considered the stuff of science fiction. We can now study the brain in "real time" when we're thinking, taking an intelligence test, practicing a craft, experiencing an emotion, or making a decision. Brain tests can even indicate when we're telling the truth, as well as provide a quick estimate of our intelligence and specific abilities.

Neuroscientists refer to this new field as cognitive science: the study of the brain mechanisms responsible for our thoughts, moods, decisions, and actions. Cognition has been defined as "the ability of the brain and nervous system to attend, identify, and act on complex stimuli." More informally, cognition refers to everything taking place in our brains that helps us to know the world. Included here are such mental activities as alertness, concentration, memory, reasoning, creativity, and emotional experience.

FROM ABNORMAL TO NORMAL TO ENHANCED

In the era of the New Brain, the emphasis is shifting from diseases and dysfunctions to an understanding of the brains of the average man and woman. An exciting consequence follows from this new emphasis on the normal brain: Research can provide us with useful guidelines about our everyday lives. For instance, recent findings indicate that by following certain brain-based guidelines anyone can achieve expert performance in sports, athletics, or academic pursuits. Such findings, of course, run counter to the traditional theory that sports achievers and geniuses are born, not made, and that our genes and other factors outside of our control impose limits on our individual capabilities. Not so. Instead, it's now clear that by learning about and applying this new research, most of us can reasonably expect greatly enhanced personal levels of achievement.

As another example, we now have good reason to believe, based on brain research, that harmful effects on our brain can result from frequent exposure to graphic scenes of violence. Moreover, it doesn't seem to matter if the violence is fictionalized, "real life," or a combination of both (i.e., docudramas featuring depictions of violence based on actual events). Watching media violence changes our brain in harmful ways that we are only beginning to understand.

I believe a lot of contemporary brain research has practical applications that can be put to use in our daily lives. Fascinating areas of cognition research include:
- Understanding the effects of media and technology on our thoughts and emotions.
- Estimating the effects of stress on brain function, with an emphasis on the use of sophisticated instruments that can help predict those people who are at greatest risk for harm.
- Formulating new brain-based ways of thinking about variations from normal behavior, such as Attention Deficit Hyperactivity Disorder (ADHD) and Obsessive Compulsive Disorder (OCD).
- Developing methods for enhancing our sensory capacities by harnessing the brain mechanisms involved in translating information from one sensory channel into another, such as the transformation of touch sensations into forms of visual perception.

Twenty-first-century discoveries about the brain will provide us with new insights into our behavior, thinking, and feelings. Thanks to technological advances, neuroscientists are already successfully correlating brain function with personality; synthesizing "designer drugs" for individualized treatments of patients suffering from depression, anxiety, and other neuropsychiatric illnesses; and correlating defective genotypes with violent or antisocial behavior.

Thanks to such advances and the promise of even greater ones in the near future, it seems fair to say that technology, rather than biology, will play the major role in the evolution of the human brain.

Unlocking the Secrets of the Brain[*]

Part I

By Tabitha M. Powledge
BioScience, June 1997

Phineas Gage may not be the most famous of the founding figures of neuroscience, but the luckless nineteenth-century American railroad builder still ranks in importance with such pioneers of brain research as Paul Broca and Carl Wernicke. Almost 150 years ago, a calamitous accidental explosion on the job rocketed a 13-pound iron rod upward into Gage's cheek, through his brain's frontal lobes, and out the top of his head, catapulting him into the annals of neuroscience.

The iron rod not only did not kill Gage, at first it did not even seem to have hurt him much, except that it cost him an eye. He moved and talked without difficulty, his memory was fine, he could still work, and his intellect appeared unaltered. But the brain damage Gage suffered did work a Jekyll-to-Hyde transformation on his personality. He changed from a kindly, cheerful, sensible, and intelligent family man, efficient and popular at work, into a profane and evil-tempered drinker, a pigheaded, willful, lazy, inconsiderate liar.

Luckily for neuroscience, his physician recorded Gage's personality changes, generating some of our earliest insights into how specialized functions are distributed in various parts of the brain. And Gage continues to contribute to neuroscience's growing body of knowledge. Scientists recently studied his skull using one of the many new methods and technologies for investigating the brain. Among the benefits of these technologies is that most of them are noninvasive—they pry open the black box of the brain, not with cranial saw and scalpel, but with mostly harmless waves from the electromagnetic spectrum. Many are mighty machines, some bulky and some sleek, that would look right at home on a Star Trek set. A few are surprisingly low-tech. But each is a revealing window on the complicated bundle of neurons that runs our lives, the organ that Nobel laureate James Watson has called "the most complex thing we have yet discovered in our universe."

Gage's brain no longer exists, but for decades his damaged skull has been exhibited at a Harvard University museum. There it drew the attention of Antonio and Hanna Damasio, of the University of Iowa Hospitals and Clinics, and several of their colleagues. In 1994, with the help of imaging studies of Gage's skull, they proposed that the rod had damaged both the left and right prefrontal cortex of the roadbuilder's brain, diminishing his ability to make rational decisions and manage his emotions.

The Damasios based their conclusions partly on a dozen of their own patients who have frontal lobe damage and suffer Gagelike defects in rationality and the processing of emotions. But to help define the exact nature of the damage to Gage, the scientists also relied on imaging techniques that pinpoint their patients' lesions. The Damasio team carefully photographed Gage's skull from several angles, measured its landmarks (including the rod's entry and exit holes), and subjected the skull to a tried-and-true imaging method: X-rays. The Damasios chose a computer image of a real brain that matched the measurements of Gage's skull, and they calculated the rod's possible trajectories through the brain. One path—through the ventral and medial sectors of the frontal lobes—fit best with both the imaging data and contemporary accounts of Gage's behavior.

The Damasio study was unusual because it employed an old but reliable imaging method, albeit updated with much fancy computer manipulation of the X-ray images. Heavy computer modeling and number-crunching is a hallmark of today's imaging techniques, which began in the 1970s with computed tomography (CT), commonly called the CAT scan, a method of "slicing" thin brain sections with X-rays and then putting them back together as computer images known as tomographs.

The Damasios also are heirs to a brain study strategy pioneered by French researcher Paul Broca beginning in 1861, the year Phineas Gage died. Broca used postmortem studies of the brains of stroke victims to trace speech defects to damage in particular brain regions. The Damasios also have specialized in learning about the brain (including its language functions) by studying people with brain damage. "We use lesions to test hypotheses about the function of large-scale systems made up of cortical regions and subcortical nuclei," Antonio Damasio says. Individual brain regions exhibit parts or components of functions, such as pronouncing a word, but complex functions, such as constructing a response to something someone else has said and expressing it as a sentence, emerge from a large network of these regions.

The lesion study method continues to work brilliantly today because researchers have amassed large patient registries—the Damasios alone have enrolled more than 2000 people—and have borrowed sophisticated neuropsychological testing tools from cognitive science. Researchers also possess space-age machines that Broca would envy, such as magnetic resonance imaging (MRI) machines, that permit them to analyze neuroanatomy in fine detail. Antonio Damasio says, "With MRI you can cut in any direction you want."

MAGNETIC RESONANCE IMAGING

MRI uses a magnetic field and radio waves to produce detailed images of brain anatomy quickly. The technique employs magnets to detect signals from the nuclei of hydrogen atoms, which consist of single protons that spin like tops. In a strong magnetic field, the protons become aligned and spin in the same direction. If the aligned protons are then zapped with radio waves at a certain frequency, the spinning hydrogen nuclei tip over and wobble. Turn off the radio wave, and the nuclei return to their upright state while emitting weak radio signals.

Scientists can deduce the amount of hydrogen in a sample by measuring the intensity of these radio signal emissions. When the "sample" is a person positioned within the immense magnetic coil of an MRI scanner, the varying concentrations of hydrogen in the body generate traces that can be analyzed by the computer and assembled into high-resolution images of the tissues and organs within. Different tissues in the brain, such as gray matter and white matter, have different chemical compositions and, therefore, different concentrations of hydrogen. Consequently, they absorb and release radio waves in different ways. The corpus callosum, which joins the two hemispheres, for example, is mainly white matter, and so it can be distinguished from the cerebellum, which is mainly gray matter. Magnetic resonance images can be three-dimensional pictures of a whole brain in any orientation or can be carved into two-dimensional "slices" of particular brain regions.

The magnetic field in an MRI machine is extremely powerful, about equal to those of the electromagnets that junkyard operators use to pick up discarded cars. This force—nearly 40,000 times as strong as Earth's own magnetic field—appears to pose no hazard to human flesh. But MRI machines can still be dangerous and even fatal. They have sucked up a hospital's floor buffers and mop buckets, transformed oxygen bottles into death-dealing missiles, and trashed both cardiac pacemakers and credit cards.

MRI, like all brain imaging techniques, has many medical applications. It can help neurologists to pinpoint brain damage shortly after a stroke. It is also the method of choice for diagnosing many other kinds of brain injuries—for example, the damage that prizefighters incur during a lifetime in the ring. MRIs have revealed specific brain abnormalities in some children exposed prenatally to alcohol, helping researchers to associate particular anatomical peculiarities (like reductions in the size of the brain, both overall and in specific regions, such as the cerebellum and basal ganglia) with cognitive symptoms such as memory deficits. And MRIs have demonstrated that the shrunken brains of adult heavy drinkers can recover lost tissue volume when the drinkers abstain.

MRI also is helping neuroscientists to sort out other kinds of brain functions that shape our lives. Elena Plante and her colleagues at the University of Arizona have been using MRI to explore the relations between neuroanatomic, behavioral, and familial components of developmental language disorder. They have discovered neuroanatomic traits that are uncommon although not grossly abnormal—

atypical volumes in certain brain regions, atypical patterns of gyri and sulci on the surface—that they believe occurred during fetal development. Plante cautions, however, that the relation between these features and a particular language disorder "is not a simple one-to-one proposition."

MRIs also have provided substantial evidence that severe emotional trauma actually damages the brain, decreasing the volume of the left and right hippocampus regions. These seahorse-shaped structures in the center of the brain are part of the limbic system and are essential in learning and memory, especially the transfer of short-term memories into permanent storage. Decrease in the size of the hippocampus shows up in adult victims of severe childhood sexual abuse as well as in veterans with combat-related post-traumatic stress disorder. These people also tend to have defects in short-term memory.

MRIs also confirm that the brains of musicians are different from other people's. The planum temporale of the left brain hemisphere, a flat and usually quarter-sized temporal lobe structure associated with the processing of sounds, is much larger in musicians than in the nonmusical—and largest of all in musicians with perfect pitch. Especially striking is the fact that the enlarged area is on the brain's left, "verbal" side, rather than the right hemisphere, long thought to be the chief seat of musical ability. The work was done at Heinrich Heine University in Düsseldorf, Germany. Senior author Gottfried Schlaug, now at Beth Israel Hospital in Boston, says that the study suggests that perfect pitch may be an innate ability, tied to development of the planum temporale at around week 30 of gestation.

Imaging techniques have also been employed for studies of the aging brain and for identifying brain differences between the sexes. MRIs have shown not only that the aging brain loses volume, but also that the patterns of loss differ in men and women. Men tend to lose volume all over the brain and in the two lobes most associated with cognitive skills, the frontal and temporal lobes. Women tend to lose more volume from the hippocampus and parietal lobes.

MAGNETIC RESONANCE IMAGING TO PROBE BRAIN FUNCTION

Standard MRI presents images of brain anatomy only, providing few clues about what is going on in the brain at a given moment. So scientists have devised several ways of peering at brain function as well as structure. Some are variations on magnetic resonance technology.

MRI observations derive from large hydrogen nuclei signals in brain water. But smaller signals from protons in other compounds and from other atomic nuclei can also be detected in living tissue, including the human brain. Study of these small signals is called magnetic resonance spectroscopy (MRS) to distinguish it from water proton MRI. MRS's virtue is that it provides chemical (and therefore physiological) information about the tissue under investigation, disclosing the presence of molecules of potassium, sodium, carbon, and other metabolites that reveal something about the state of the tissue. MRS has been in use for more than a decade, exploring brain biochemistry in schizophrenia and bipolar disorder; helping to plan brain surgery; and diagnosing coma, dementia, oxygen deprivation in newborns, stroke, and head injury.

Conventional MRI can detect brain tumors, but a biopsy is usually necessary to determine a tumor's type and whether it is malignant. By contrast, MRS can distinguish between different types of brain tumors, a technique that could enable patients to avoid sometimes risky biopsies. Using noninvasive MRS, Douglas Arnold and his colleagues at McGill University were able to classify 90 out of 91 tumors correctly because different tumor types have different characteristic metabolite patterns.

One disadvantage of MRS is that finding the tiny chemical signals usually requires minutes rather than the seconds a standard MRI takes. But MRS's ability to measure biochemistry directly in the human brain is unique. In addition, it can be combined with the anatomical images generated by conventional MRI to yield additional useful details. The two can even be done in the same machine.

The newest, and perhaps brightest, star in the cluster of magnetic resonance techniques is functional magnetic resonance imaging (fMRI), in which the radio signals from protons increase when the level of blood oxygen goes up in particular regions of the brain. Oxygen level is an index of those regions' activity—that is, of the brain's response to thought and movement. Researchers can "scroll" through fMRI scans on a computer monitor or videotape, obtaining evidence about brain activity over time.

Like all the brain imaging methods, fMRI has many medical applications, such as providing guidance to brain surgeons. But, like the others, it is also used to map normal brain functions, such as how visual stimuli are processed and how tastes are perceived. Some experts expect fMRI to be the dominant tool for brain mapping for the next several years.

Thanks to fMRI, scientists may have located the overseer region that supervises nervous impulses in and out of a kind of buffer system called working memory, where information is temporarily stored and manipulated. They have hypothesized that working memory (often referred to as short-term, or telephone number, memory) must possess what they call a "central executive system" (CES) that acts as traffic cop. To find this system, Mark D'Esposito and his colleagues at the University of Pennsylvania Medical Center gave their study subjects two easy tasks to perform. One was to identify vegetables on a list of miscellaneous words; the other was to specify which of two squares contained a dot in the same location as a target square that had been rotated in space.

When subjects carried out the assignments separately, fMRI revealed activation in the left temporal lobe for the semantic task and in the parietal and occipital regions in the back of the brain for the dot-location task. When subjects had to perform both tasks at the same time, shuttling data in and out of short-term memory, the prefrontal cortex in the front of the brain also lit up, along with the front of the cingulate gyrus, a C-shaped component of the limbic system. "[O]ur findings support the hypothesis that dorsolateral prefrontal cortex is involved in the allocation and coordination of attentional resources," the researchers wrote in the 16 November 1995 issue of *Nature*. "Moreover, recruitment of anterior cingulate and dorsolateral prefrontal cortex, which are anatomically interconnected, suggests that the CES may comprise several components."

Evidence obtained with this technique is also persuading scientists to revise some long-held beliefs about what various parts of the brain do. Using fMRI, researchers have confirmed other studies suggesting that the cerebellum, the wrinkled ball of tissue about the size of a fist that snuggles under the cortex, does more than just manage movement, the job that neuroscientists classically attributed to it. The cerebellum also appears to handle at least some sensory information, such as the texture and shape of objects. The cerebella of subjects undergoing fMRI were at their liveliest when the subjects were trying to distinguish between types of sandpaper and the shapes of small balls simply by feeling them with their fingers.

Functional MRI also has pinpointed the site of the human "interval timer," which helps us keep track of short periods of time. Warren Meck of Duke University reports that it resides in the striatum, deep within the brain. He and his colleagues found it by asking volunteers to squeeze a ball every 11 seconds and watching to see which part of the brain lit up during the exercise.

Imaging systems are a favorite tool for studying how language relates to the brain, and fMRI is no exception. It has, for example, provided support for the growing suspicion that what underlies dyslexia—problems with reading, writing, and spelling—may be defects in timing ability, not just language.

Using fMRI, Guinevere Eden and her colleagues at the National Institute of Mental Health have found that dyslexic men are somewhat less able to detect visual motion than are nondyslexic men. These men also fail to activate the brain region that responds most strongly to visual motion. Known as V5, it is located in the cortex at the junction of the occipital and temporal lobes. The researchers speculated in the 4 July 1995 issue of *Nature* that this subtle defect in perceiving visual motion actually indicates a more general deficit in timing ability. The deficit could include relative insensitivity to rapid changes in the auditory system, which might explain, for example, why dyslexics find it hard to hear the differences between consonants. The researchers also suggest that V5 inactivation might serve as an early biological marker for dyslexia that is completely unrelated to reading itself. This finding might make it possible to spot potential dyslexics before first grade and so stave off school failures.

Although fMRI is the most fashionable imaging technique right now, several additional ways of looking at the brain also possess distinct advantages. Like fMRI, many of them permit scientists to study the living brain at work. Part II of this article will look at six more of these techniques.

Unlocking the Secrets of the Brain*

Part II

By Tabitha M. Powledge
BioScience, July/August 1997

Functional magnetic resonance imaging, described in Part I of this two-part series, is the imaging experts' darling of the moment. But when the rest of us hear the term "brain imaging," we mostly think of the dazzling colored pictures of the brain that we see everywhere—in slick magazines, TV commercials, and print ads, on the Web, and even (although they are usually fuzzy and muddy) in newspapers. Those eye-popping images are generated by positron emission tomography (PET).

POSITRON EMISSION TOMOGRAPHY

PET was the first scanning method to yield information about brain function rather than simply anatomy. PET measures concentrations of positron-emitting radioisotopes in living tissue, recording signals from radioactive tracers that researchers administer to the study subject. A computer then analyzes the patterns of radioactivity, turning them into colorful cross-sectional pictures that depict biochemical events as they occur during the action under study. PET scanners have measured glucose metabolism, cerebral blood flow, oxygen metabolism, dopamine, opiates, serotonin, and glutamate, among other physiological molecules and processes.

Many compounds can serve as radioactive tracers, but most tracers are short-lived radioisotopes of elements that occur commonly in the human body: carbon, nitrogen, oxygen, and fluorine. Isotopes with short half-lives—in many cases only a few minutes, and rarely more than an hour or two—reduce the administrators' and the subjects' risk from exposure to radioactivity. One limitation of this technique is that same brief half-life, which requires on-site production of the radioisotopes. Onsite production can be accomplished at institutions staffed with nuclear physicists, computer experts, and radiopharmacists who have access to cyclotrons. But most institutions lack such facilities, which explains why only 100 or so PET-scanning centers exist worldwide, most of them dedicated to research.

Single photon emission computed tomography (SPECT) is similar to PET in that it images blood flow in the brain and also uses radioactive tracers, but it detects a different type of photon. Its images are comparatively low resolution, but it uses isotopes with much longer half-lives than PET does, so the isotopes do not require on-site production. This simpler supply of isotopes is one reason that SPECT is much less expensive than PET and also more often available in diagnostic centers.

SPECT is often used to study brain dopamine and has, for example, corroborated results from postmortem studies that suggest that dysregulation in presynaptic dopamine function may be responsible for Tourette's syndrome. Subsequent SPECT studies in twins have established that genetically identical Tourette patients can differ in the severity of their symptoms because of differences in dopamine receptors. A team led by Daniel Weinberger of the National Institute of Mental Health (NIMH) in Rockville, Maryland, used SPECT to scan the brains of twins for up to four hours and found that binding to dopamine receptors was higher in those whose tics and other symptoms were worse, although why identical twins differ remains a mystery.

The uses to which researchers put PET is just about endless, but the best known may be studies of language. The now-classic PET images made by Marcus Raichle, of the Washington University School of Medicine, show changes in local blood flow associated with local changes in neuron activity that occur during different types of information processing. Hearing words, seeing words, speaking words, and generating words (the mental activity that occurs when people think about words before saying them) show up as splotches of red, yellow, green, and blue in the different parts of the brain that carry out each of these tasks. Raichle's PET pictures, perhaps more than any others, made the case for the decentralized brain and carried that message to scientists and the public alike.

All imaging methods reinforce the idea that the brain is organized in modules, although they also have indicated clearly that the modules are flexible rather than rigid. No imaging method has demonstrated this better than PET. Especially persuasive evidence of the flexibility of brain modules comes from PET studies of people with clinical disorders in language, such as loss of knowledge about a specific category of objects as a result of brain damage. Some patients may be unable to identify or name living things; others struggle to name human-made objects. A group at NIMH, led by James Haxby, has found that the brain region that becomes active during the naming of objects varies depending on the type of objects involved.

For example, naming animals selectively activates the left medial occipital lobe, a brain region involved in the earliest stages of visual processing. Naming tools activates both a left premotor area that also is activated by imagined hand movements and an area in the left middle temporal gyrus that also lights up when the subject is generating action words. Both kinds of naming activate the ventral temporal lobes on both sides of the brain. Haxby's group concluded that semantic representations of objects are stored as a distributed neural network that includes the ventral region of the temporal lobe plus other areas whose participation depends on the properties of the object to be identified.

The idea that the brain is modular, but flexibly so, has been reinforced by the work of Hanna and Antonio Damasio of the University of Iowa, after Raichle probably the researchers best known for images of how the brain handles language. In the 11 April 1996 issue of *Nature*, they proposed a novel view of the brain's storehouse of words. For some time, researchers have believed not only that our knowledge of words is organized by category, but also that the brain contains representations of words that are separate from both the meaning and sound of the words. Drawing partly on PET studies of people who show no evidence of impaired language skills, the Damasios have argued that these lexical representations lie in several networked regions of the left hemisphere that are outside the specific language areas in the left temporal and parietal lobes.

"Depending on the kind of concept you are dealing with, what you have in the brain is a dictionary access arrangement," says Antonio Damasio. He and his colleagues are suggesting, he said, that the dictionary is dynamic and linked to each person's experience. "Our knowledge is built on bits and pieces of many aspects of a given thing—shape, color, movement, taste. Those things are not going to be laid down in one single place," he said. Thus your concept of, say, a cat is decentralized, composed of a number of kitty qualities—warm and fuzzy, four legs and whiskers, teeth and claws, meows and purrs.

In addition to helping scientists to understand brain architecture, PET also has helped scientists to tackle social problems and social policy questions. Researchers are identifying the brain mechanisms that operate in drug addiction and have found that some pathways are employed by more than one addictive substance. In fact, addicting drugs in general seem to activate a single pleasure circuit—the one for the neurotransmitter dopamine, formerly called the "reward center"—in the oldest part of the brain.

PET scans have helped demonstrate for the first time that the human brain's own natural opiate system is involved in cocaine dependence and cocaine craving, as reported in the November 1996 issue of *Nature Medicine*. Consistent use of cocaine causes the brain to lessen production of its own opiates, the enkephalins, and simultaneously to increase the number of opiate receptors, which allows the brain to scoop up as many enkephalins as possible from the reduced supply. PET scans showed that a nonaddicting synthetic drug that binds to opiate receptors increased its binding in addicts compared with controls. James Frost, of Johns Hopkins University, and his colleagues proposed that craving results when a large number of the extra opiate receptors that cocaine creates remain vacant. They plan to study how other addictive drugs, such as nicotine and alcohol, interact with opiate receptors.

PET has also shed light on how individual drugs do their work—for example, how amphetamines heighten alertness and sharpen thinking. Researchers at NIMH have found that the drugs work by enhancing only the brain areas that are essential for a specific job. Simple tests of abstract reasoning normally activate the prefrontal cortex, whereas more complex problem-solving activities center on the hippocampus. Subjects receiving small amounts of amphetamines showed three effects: an even greater increase in prefrontal activity during an abstract-reasoning task, a similar hippocampal increase during problem solving, and large reductions in activity in parts of the brain that are not normally active during such tasks. "The amphetamines respected the normal landscape by enhancing activity where it was appropriate and decreasing it where it was not," said NIMH's Daniel Weinberger. "This is probably what attention is all about."

PET also has helped scientists to venture into philosophy, exploring an ancient topic that philosophers call the "mind-body problem." In its simplest form, the mind-body problem asks about the relationship between mental activities, such as thinking and imaging, and physical events taking place in the brain.

PET, for example, has revealed that psychotherapy—"the talking cure"—affects brain function physically, not just psychologically. Moreover, these biological effects are remarkably similar to those achieved by purely physiological therapy, which employs only psychoactive drugs. The seminal study, published in the 15 February 1996 issue of *The Archives of General Psychiatry* by a team at the University of California at Los Angeles, reported that ten weeks of cognitive therapy not only produced a substantial improvement in six people with obsessive-compulsive disorder but also eased the patients' abnormally tight links among four brain areas (the orbital frontal cortex, the caudate nucleus, the cingulate gyrus, and the thalamus)—links believed to underlie the disorder. PET revealed that therapy apparently permitted the four areas to work more independently of one another. Effects on the caudate nucleus in particular were similar to those wrought by drugs effective against the disorder, such as Prozac.

That conversation can change the brain should perhaps not be too surprising. We now know that our brains remodel themselves all the time, creating new synapses, growing new dendrites, forging new connections among neurons as they absorb new experiences and create new memories. After all, cognitive behavioral therapy is just another form of learning. Following instruction, patients changed their brains, and their behavior, by making a conscious effort to resist compulsions such as the desire to wash their hands constantly. By continually telling themselves that the behavior was just a compulsive urge, the unnecessary product of a medical condition, they learned to apply a different label to the urges. As a result, the urges lessened.

Perhaps the most intriguing recent revelation from a PET scanner is that the brain seems able to distinguish false memories from true ones. In a telling experiment, Daniel Schacter, of Harvard University, and his colleagues read 24 lists of

20 words to each of 12 women. Ten minutes later, the researchers gave them somewhat different written lists and asked which words in the second group were also in the first. The women were almost as likely to pick words that they had not heard as words that they had.

The subjects were PET scanned during recall, and, as expected, the hippocampus—long believed to be a center for memory retrieval—lit up whether they were recalling a word that they actually had heard or one that they only thought they had heard. But when they were recalling words that they actually had heard, the scan showed activation in another place, too—the left temporal parietal area, where sounds are decoded. The true memory had apparently retrieved not just the word, but also a sensory detail—the actual sound of the word—that had been laid down at the instant of learning. When a subject recalled a word erroneously, only the hippocampus lit up.

This work may carry important implications for the criminal justice system, because some observers believe that it could lead to the development of a reliable lie-detector test. Schacter, however, downplays this potential. For one thing, his study subjects were quizzed within ten minutes of seeing the data. No one knows whether the findings would be similar if taken months or years later. Sensory details are likely to fade with time, so relevant sensory areas might not light up, even for a true memory. People also may eventually attach fictional sensory details to false memories, making them seem truer.

ELECTROENCEPHALOGRAPHY

"Brain waves don't get very much play these days in terms of modern neuroimaging technology," Alan Gevins observes a little sadly. "One could say a lot about why that's so. I think it's mostly because [brainwave technology is] old." Gevins runs EEG Systems Laboratory, a private lab in San Francisco that he spun off from the University of California-San Francisco some years ago. Quite likely, no one knows more about brain waves than he does.

An electroencephalogram (EEG) simply measures the amount and type of electrical activity in the brain. The brain's electrical impulses are detected by electrodes placed on the scalp, amplified approximately a million times, and recorded on moving graph paper as wavy lines—hence, brain waves. EEGs are often used to diagnose certain illnesses, especially seizure disorders but also brain tumors, head injuries, degenerative diseases, and others. EEG is also the definitive test for brain death.

As a research tool, the study of brain waves has a long and occasionally checkered history, but it also has some big advantages. First, and perhaps most important, it is the only method of looking at the brain that yields information in real time—the temporal resolution of an EEG is measured in milliseconds.

Second, EEG technology is dirtcheap compared with the big new scanners and all their impedimenta, like cyclotrons. "A whole brain-wave system costs one to two orders of magnitude less than PET or MRI scanners," Gevins says.

Third, brain-wave technology is portable. Gevins has recorded the brain waves of people as they walk, talk, and otherwise go about everyday activities as well as the brain waves of race-car drivers moving at 100 miles per hour and of stunt pilots doing aerobatics hundreds of feet up. "I can record brain waves from anyone anywhere, just about," Gevins claims. "For the next 10 or 15 years at least that's going to be a unique and valuable feature of brain-wave studies. You can measure brain function in the real world rather than in a brain scanner with a subject immobilized."

And the bad news? "It's not a true 3-D imaging modality," Gevins says. "It's maybe 2 ½-D. I can get a nice image of activity on the superficial surface of the cortex, and for some applications that's sufficient."

The field's pioneer was German psychiatrist Hans Berger, who measured brain waves in his daughter as she was doing mental arithmetic. He noticed that a large amplitude potential of her brain became small when she tried to multiply 5 ⅛ by 3 ⅓ squared. When she finished the problem, the large-amplitude wave—which came to be called the alpha rhythm—returned. "This was a seminal observation, made in the late 1920s, which directly related a measurement of electrical brain function to mental activity," Gevins says. Since the 1920s, scientists have discovered a number of other characteristic human brain waves: beta (between 13 and 30 cycles per second, normally a sign of alertness and attention); delta (less than 4 cycles per second, occurring normally during sleep and in young children); and theta (from 4 to 8 cycles per second; excess theta may indicate brain injury).

In the 1960s, researchers discovered that they could record brain waves in two ways, either as continuous, ongoing activity, recorded literally for days at a time, or as momentary activity, correlated to the presentation of a discrete stimulus, for example, a flash card demanding that the subject add two numbers. When the latter is done many times, and researchers average the brain waves and time-register them to the stimulus, the tracings are called "evoked potentials" or, sometimes, "event-related potentials" (ERPs).

Researchers in the 1960s were interested in making topographic maps of brain activity as measured on the scalp and then trying to relate these patterns to the then-fashionable notion that the right hemisphere is specialized for spatial, holistic functions and the left hemisphere for language. "This was a very popular sport in the late 1970s and early 1980s," Gevins recalls. "Religious cults formed around one hemisphere or the other." Thousands of studies were published, Gevins says, "but many of them weren't so good." Many failed to control stimulus and response activity or to isolate the mental function to be measured. "But one really important thing that came out of all those studies was the knowledge of how to do experiments, how to isolate perceptual factors and motoric factors and manipulate mental factors," Gevins says.

Throughout the 1980s, Gevins's lab and others tried to develop measurements of brain waves that would characterize distributed functional networks in the brain quickly. "A thought occurs"—Gevins snaps his fingers—"in a fraction of a second. In that fraction of a second, many processes take place: perceiving the stimulus, breaking it down into its pieces, putting the pieces together, deciding what it means, preparing to execute an action, executing the action. We'd like to be able to break that fraction of a second into little pieces and say something about the distribution of activity in the cerebral cortex during this single period of time."

Gevins has also discovered that as subjects learn a task, they concentrate more efficiently, the task becomes easier, and alpha waves recorded over the backs of their heads get bigger and bigger. "These signals of focused attention and mental effort change nicely and very predictably with the difficulty of the task, and they differ between people of average and superior intelligence."

Gevins suggests that monitoring brain waves might be useful in a number of clinical settings—for example, tracking how a psychotropic drug affects a patient's ability to concentrate or how stroke rehabilitation treatment is coming along. Among the best-known EEG studies are those reporting unusual brainwave patterns in male alcoholics. At first, these patterns were assumed to be due to the toxic effects of alcohol, but they also have turned up in the young sons of male alcoholics long before the boys have had a drink. The father-son patterns are now interpreted as further evidence of an inherited predisposition to alcoholism (and perhaps to abuse of other drugs) and may offer a way of detecting youngsters at risk.

The EEG has also proved superior to behavioral measures like reaction time, because ERPs can reveal signs of neural processing well before the body responds with motor output. A team at the Centre de Recherche Cerveau et Cognition in Toulouse, France, called on ERPs for studying how long it takes the human visual system to process a complex natural image. The researchers asked 15 subjects whether a photo they saw for only 20 milliseconds contained an animal. The subjects' responses averaged 94% accuracy, and the ERPs indicated that the brain's processing of the visual information was achieved in under 150 milliseconds.

Gevins has suggested that ERPs' capacity to foreshadow later performance based on subtle, early electrical signals could be harnessed to make alertness monitors. Some years ago, he attached his electrodes to the scalps of a handful of Air Force fighter test pilots and subjected them to many hours of boring but difficult tests of manual dexterity. Hours before their performance deteriorated, changes in their brain-wave patterns foreshadowed the decline of performance.

LOW-TECH WAYS OF LOOKING AT A BRAIN

The most ancient way of looking at a brain—cutting it open after its owner is beyond objecting—is still in wide use for research. Scientists have employed postmortem studies, for example, to examine thorny questions of sex roles, sex differences, and sexual orientation. The best-known such study was the controversial report in the 30 August 1991 *Science* by Simon LeVay, then an associate professor at the Salk Institute, which showed that the brains of heterosexual and homosexual men differed in a small portion of the preoptic area.

In the 2 November 1995 issue of *Nature*, scientists at the Netherlands Institute for Brain Research in Amsterdam published their conclusion that there is a structural difference between the brains of ordinary men and male-to-female transsexuals. The difference shows up in a tiny portion of the hypothalamus, which is smaller in women and in male-to-female transsexuals than in men. Whatever the cause—the researchers speculate that it is due to aberrant hormone levels during fetal life—the difference appears to have nothing to do with sexual orientation, because the brains of homosexual and heterosexual men do not differ in this way. The area instead seems linked to body image.

In a commentary on the Dutch study in the same issue of *Nature*, S. Marc Breedlove, of the University of California-Berkeley, contended that the research presented exactly the same dilemma that LeVay's did: Are these brain differences genetic (or at least fetal) in origin, or do they result from life's experiences? "Laymen may assume that a structural difference in the brain is the immutable signature of purely biological forces, but three decades of neuroscience research have made it clear that experience can dramatically alter the structure and function of the brain," Breedlove wrote. "At present, these brain regions can only be measured postmortem. Until technology enables us to measure them repeatedly in the same person at different ages (before and after puberty, for example), there can be no definitive answer as to whether these regions direct psychological sexual differentiation or are themselves directed by that process."

Vilayanur Ramachandran, of the University of California-San Diego, eschews scanners and even scalpels in his low-tech approach to examining brain function. Resorting to tools such as ordinary cotton swabs, he makes progress on such perplexing questions as phantom-limb pain, long a source of embarrassment and anguish to patients and enormous frustration to their physicians. Ramachandran asks amputees to close their eyes, then strokes their faces with cotton swabs. They "feel" the swabs on their nonexistent arms, leading the scientist to conclude that the sensory cortex has reorganized itself so that its maps of arms and faces now mingle.

PEERING INTO THE FUTURE OF LOOKING AT A BRAIN

Researchers have grown increasingly interested in melding together different imaging methods so that they can draw on the strengths of each to yield more informative pictures of just what the brain is up to. Scott Lukas, of Harvard Medical School, has combined EEG with magnetic resonance imaging (MRI) to produce a kind of topographic brain map showing that cocaine increases alpha waves. Linda Chang, of Harbor-UCLA Medical Center in Torrance, California, is combining

MRI, magnetic resonance spectroscopy, and SPECT to study both the effects of cocaine on AIDS dementia and the neurotoxicity of the drug known as ecstasy.

Magnetoencephalography is similar to EEG except that it measures changes in magnetic fields created by the brain's electrical currents. Researchers combined it with MRI to produce a three-dimensional map of the brain areas activated by touching the fingers of one hand. Rodolfo Llinás and his colleagues at New York University found that this map was distorted in a patient born with webbed fingers. Within a few weeks after surgeons separated the fingers, however, the patient's brain map had transformed itself into something approaching normal.

Of course, present techniques will be refined and improved. A late-model MRI machine accommodates two surgeons as well as the patient, so that live, three-dimensional images can guide an operation. And new techniques are always in development. A powerful pulsing magnet is central to transcranial magnetic stimulation, an experimental way of mapping the brain that is said to be capable of causing muscles to move and moods to change dramatically. The pulses paralyze tiny areas of the brain, producing temporary lesions that affect its function and allow scientists to locate the physical site of sensations, feelings, and frames of mind.

Many brain studies—especially those based on imaging, which has rendered the brain as essentially a large cluster of specialized modules that each carry out a specific task—may seem, at first, to reinforce the notion of extreme localization of brain function. But that is not what the images are telling us, experts say. "Rather, it's the discovery of groups of areas that perform observable human behavior," says Raichle.

Imaging has helped to mold today's consensus among neuroscientists that many brain functions are distributed around versatile, adaptable networks with nodes in several places and that groups of neurons can be recruited—often quite easily—to new tasks. "You can think of it in terms of a large symphony orchestra but with a finite number of players who have specific roles to play in all of this," Raichle says. Even with a finite number of players, he points out, an orchestra can generate an infinite number of brilliantly coordinated performances.

Of course, a metaphor involving the brain as an orchestra raises the issue of who is conducting. "I don't know," Raichle says. "Many fine orchestras can perform without a conductor. I happen to be a symphony musician. I've generally felt the conductor had an important role, but it is also true that a well-honed orchestra can play without one. Where is the conductor...I'm not going to pose an answer for you other than to say that it's obviously occurring."

And so, armed with an increasing number of ways, both fancy and plain, for looking at the brain, neuroscientists plan to keep searching.

MRI: A Window on the Brain*

By Paul Raeburn

Technology Review (Cambridge, Mass.), December 2005/January 2006

When Bradley Peterson, a psychiatrist and researcher at Columbia University, offered to scan my brain with a magnetic resonance imager the size of a small Airstream trailer, I immediately said yes. I spent 10 minutes filling out a page-long checklist (I lied on the question asking whether I was claustrophobic) and another few minutes emptying my pockets and getting rid of keys, wristwatch, and pen, which could become missiles inside the MRI's potent magnetic field.

I lay down on a narrow pallet that slid into the machine like a drawer in a morgue. The machine groaned and clanged as it peered inside my skull, then fell silent. With a gentle whir, the pallet slid out, and I relaxed. In about the time it takes to burn a few CDs on my laptop, Peterson was leaning over a screen, showing me a detailed black-and-white image of my brain.

Brain scans like the one I had are now routine, used for everything from detecting signs of stroke to searching out suspected tumors. But researchers like Peterson are pushing MRI technology further than anyone once thought it could go. In the last decade or so, MRI has been retooled to reveal not only the anatomy of the brain but also the way the brain works.

While conventional MRI scans, like the one Peterson gave me, reveal physiological structures, a variation called functional MRI (fMRI) can now also image blood flow over time, allowing researchers to see which areas of the brain are active during certain tasks.

Indeed, fMRI studies over the last few years have provided researchers with startling images of the brain actually at work. A yet newer extension is MRI spectroscopy, another kind of functional imaging that monitors the activity of particular chemicals in the brain—providing different clues to brain function than fMRI does. And most recently, researchers have pioneered an MRI technique called diffusion tensor imaging (DTI) that produces 3-D images of the frail, spidery network of wires that connects one part of the brain to another.

MRI has become, says Robert Desimone, director of the McGovern Institute for Brain Research at MIT, "the most powerful tool for studying the human brain. I liken it to the invention of the telescope for astronomers." Desimone notes that the arrival of the telescope did not immediately revolutionize the scientific understanding of the universe. That took time, as researchers learned how to use their new tool.

The same thing is happening with MRI, Desimone says. Researchers are just now beginning to realize the potential of these techniques, which were first widely used on humans about 15 years ago. "You're seeing a lot of excitement in the field," says Desimone.

Several technical advances have contributed to MRI's improvement. Topping the list is the development of more-powerful MRI magnets, which enable more-detailed, higher-resolution scans. What megapixels are for a digital camera, teslas, a measure of magnetic-field strength, are for MRIs: the more you have, the better the quality of the image. The newest MRIs generate magnetic fields of about seven teslas, many thousands of times stronger than Earth's magnetic field and at least twice as strong as those typically used in hospitals. (Some research centers, including the McGovern Institute, have 9.4-tesla MRI scanners for animal studies.)

Another key development is a succession of ever more complex methods of computer analysis. These allow researchers to extract more and better information from scanner data and have improved not just fMRI but also MRI spectroscopy and DTI.

The ultimate aim of brain imaging research is to help explain how the billions of neurons and connections in the brain give rise to thought. But researchers are also applying the new MRI techniques to a more practical, immediate goal: improving the diagnosis and treatment of mental illnesses and learning disorders. The hope is that MRI imaging will provide far more accurate diagnosis of psychiatric diseases whose symptoms can resemble each other, preventing years of suffering for patients put on the wrong medications.

As part of this effort, researchers are using MRI to investigate the causes not only of psychiatric ailments but of all kinds of brain abnormalities and learning disorders, including those often found in children born prematurely. And while attempts to use brain imaging to improve psychiatric health care have met with little success over the last decade, the new MRI technologies—in essence, far stronger telescopes on the mind—are providing fresh hope of finding better ways to intervene.

BIPOLAR FINGERPRINT

One of the leaders in the effort to enlist MRI in the diagnosis and treatment of psychiatric ailments is John Port at the Mayo Clinic in Rochester, MN. Port is a neuroradiologist who began his career by studying electrical engineering and computer science at MIT and later earned a PhD in cell biology and an MD from the University of Illinois. So he's in a good position to research both basic MRI technology and its applications to medicine.

Port's work on MRI could have broad application in psychiatry, but for now he is concentrating on his particular interest: bipolar disorder. Also called manic-depression, bipolar disorder is characterized by mood swings from wild exuberance to profound depression, with periods of stability in between. X-rays or conventional MRIs show no difference between the brains of people with bipolar disorder and those without it; medical journals are littered with failed attempts to use imaging to find distinctive signs of the disease.

Port thinks a lot of those attempts were scientifically flawed. "I have a list of pet peeves a mile long," he says. "There are a million studies, but the patients might be on six different medications. So when you see something different, is it the meds? Or is something going on?" Another problem with many earlier studies, he says, is that they included too few patients. "You can't tell anything from 10 patients. A lot of the research hasn't been as rigorous as it should be."

Indeed, despite years of work, neuroscientists still do not know what causes bipolar disorder, or exactly which parts of the brain are involved. That lack of knowledge has severely hampered the search for safer and more effective ways to treat the disease. The principal drugs for bipolar disorder, lithium and Depakote, have been around for decades.

Both were discovered by accident, when researchers trying to do something else noticed that the drugs eased the symptoms of patients with bipolar disorder. And though the drugs can be reasonably effective in some people, doctors have no idea how they work or which patients are most likely to benefit. In order to find better pharmaceuticals, researchers need to be able to target the exact mechanisms or structures involved in bipolar disorder.

Pinpointing the mechanisms could also lead to more accurate evaluation of the disorder. Often, diagnosis in psychiatry is done by a kind of trial and error, in which a psychiatrist makes an educated guess based on the behavior or self-reported symptoms of a patient, prescribes a medication, and sees whether or not it helps. If it doesn't, the psychiatrist considers a different diagnosis and a different medication, until something begins to work.

"What psychiatrists need is some test that will give them the answer: this patient has the disease or doesn't," says Port. He and other researchers hope MRI scanners will offer the definitive diagnosis. And for those in the mental-health profession, that would change everything. "I'm dedicating the rest of my career to coming up with an imaging test that will help psychiatrists diagnose" bipolar disorder and other illnesses, Port says.

Port is one of many researchers now experimenting with MRI spectroscopy, in which software produces an image of the brain based on a spectroscopic scan. The image is made up of individual data points called voxels, cubes analogous to the pixels in a 2-D computer image. Each corresponds to a volume about the size of a kidney bean. For each voxel, Port gets a reading on the presence or absence of certain chemicals that are indicators of brain function.

To understand how MRI spectroscopy works, it's necessary to understand a bit about how magnetic resonance imaging works more generally. MRI scanners pick up extremely faint electromagnetic signals coming from protons in the atoms of molecules that make up the body's tissues—in this case, brain tissue.

"Think of it like listening for a pin drop in a thunderstorm," Port says. Each proton has a magnetic field that points in a certain direction, as the earth's does. When the MRI is turned on, its magnet aligns the protons' magnetic fields in the same direction. Bursts of radio frequency energy temporarily knock some of the protons out of alignment. When the protons snap back into place, they release energy, generating a minuscule signal that the MRI's detectors can pick up. By flipping the protons different ways and measuring various properties of those flips, including the time they take, researchers can identify various tissues and chemicals in the brain.

Using MRI spectroscopy, Port can measure levels of chemicals such as n-acetyl aspartate, which is found only in neurons, or glutamate, which stimulates nerve-cell activity. When Port used the technique across many areas of the brain in bipolar patients and compared the results to those from healthy controls, he came up with a chemical fingerprint that seemed to be an indicator of bipolar disorder.

"When we compared all the bipolar patients in any mood state with their matched normal control subjects, we found that two areas of the brain were significantly different," Port says. Port and his team also identified changes in many regions of the brains of people with bipolar disorder that indicated whether they were in a manic state or depressed. "We found a chemical measure of the mood state," he says.

So has Port found the long-sought diagnostic test for bipolar disorder? Does his chemical fingerprint reliably identify people who have bipolar disorder and exclude those who don't?

Maybe, but he can't be sure yet. "We think we're on to something good," he says, but "we have to check it and make sure it will be clinically useful." It's a question of trying the technique with enough patients to be sure that it is statistically valid—that it won't produce too many false positives or false negatives. It doesn't have to be perfect, but it has to be good enough to add useful information to what psychiatrists can discern through their traditional methods of diagnosis, interviews, and analyses of patient histories.

If Port is correct, however, and the technique proves itself, it would be a landmark in psychiatric research: a diagnostic test for bipolar disorder. And if the technique works with bipolar disorder, it could be adaptable to other psychiatric illnesses.

Port and others are also experimenting with diffusion tensor imaging. DTI measures water diffusion in the brain. Water flows through the brain as it does anywhere else—along the path of least resistance. In the brain, that's along the axons, the neurons' long tails, which convey electrical signals to other neurons. (It's from the fatty, white insulation that surrounds most axons that "white matter" takes its name; the rest of the neuron, and uninsulated axons, together constitute "gray matter.")

Port is just beginning to research the technique. But eventually researchers will be able to use "DTI clinically to look for diseases that interfere with white matter—amyotrophic lateral sclerosis [Lou Gehrig's disease] and schizophrenia," Port says.

DIAGNOSING DEVELOPMENT

The techniques Port is studying, if they prove successful, will be used in diagnosing people already showing signs of mental illness. But what about others who are predisposed to problems but have not yet begun to exhibit symptoms? Can the MRI technology help to find these people so that they can be helped before symptoms appear?

At Columbia, Peterson is trying to answer that question. He and collaborators are among the first to scan the brains of premature infants—sometimes within days of their birth. The aim is to catalogue the types of brain abnormalities they discover and to devise ways to intervene earlier than ever before to try to correct or compensate for them.

Peterson first became interested in the complications of premature birth about 10 years ago, when he was beginning his psychiatric research at Yale University. He had discovered something very unusual in the brains of people with Tourette's syndrome. Most of us have asymmetries in our brains—the left side doesn't exactly match the right. Most of us also have one eye that's bigger than the other (as portrait photographers will point out) and other minor asymmetries.

But the brains of people with Tourette's syndrome were different. "In the Tourette's brain, there seemed to be an absence of asymmetry," Peterson says. A similar absence of asymmetry had been observed in animals that survived complicated births. Peterson decided to look at children who had been born prematurely. Like Port, he is using the newest MRI technologies to try to obtain information that hasn't been available before.

There was a reason for his interest. Children born prematurely are at greater risk for learning disabilities and even psychiatric illnesses. Understanding how their brains are different should lead to new ways to help them.

As it happened, Laura Rowe Ment, a pediatric neurologist at Yale, was following a group of 500 premature children born between 1989 and 1992 as part of an ongoing study. Peterson and Ment set up a collaboration. "There were imaging reports suggesting various kinds of problems in the brain—in terms of brain development. But they were uncontrolled, the numbers were small—they were impressionistic," says Peterson.

Even given their smaller body size, premature kids tend to have disproportionately small heads. "The guess was that brain size would be reduced" later in life, says Peterson. Researchers also speculated that there would be damage to the white matter. Ment's kids, who were then about eight years old, were especially useful because she and her colleagues had documented everything that had happened to them since they were born.

The first thing Peterson did was use the MRI scanner to determine the size of the eight-year-old children's brains. The guess was right—their brains were smaller than normal. But the decrease in size occurred only in certain brain regions—the parts of the cortex that govern movement, vision, language, memory, and visual and spatial reasoning. "These regions were dramatically smaller," Peterson says. The other parts of their brains were normal size, or close to it.

The second guess—about damage to white matter—also proved accurate. There was less white matter in the motor regions of the children's brains, meaning there were relatively few wiring connections there. And the reduction in volume correlated with IQ scores. "The bigger the abnormality—the more abnormal it was in all these regions—the lower their IQ was," Peterson says.

The question then was, Did these abnormalities arise at or before birth or sometime later? Peterson started scanning normal and premature infants. The scans of premature newborns showed that they had the same brain abnormalities as the eight-year-olds. "It was so distinctive, the pattern of abnormalities, it's almost impossible to look at a scan and not be able to tell this is a premature child," Peterson says.

One of the most salient differences was in the size of the tiny cavities in the brain known as ventricles. "The ventricles are massively dilated, about four times larger in the prematurely born kids than in the term children," Peterson says. "We saw that in eight-year-olds and in the infants. The tissue around those ventricles is really damaged. . . . It suggests that these babies are having problems in development even before they're born." Peterson followed the newborns for two years and then tested them with a kind of IQ test meant for toddlers. The earlier they were born, the more immature their brains were at birth. And the more immature their brains, the lower their intelligence scores.

To neuroscientists, the discovery that premature kids had brain abnormalities made sense. Much of the brain's growth and development occurs during the last half of pregnancy. Neurons begin life clumped near the center of what will become the brain but soon start to migrate outward. Glial cells, which help neurons communicate, go through a period of explosive growth, accounting for most of the brain's increase in weight. The neurons extend meandering tentacles, seeking connections with other cells. Billions of connections are made during the last weeks of pregnancy. The axons then develop their coats of white, fatty insulation. By this time, the brain is massively overdeveloped, with far too many wires and connections. So it begins cutting back. It's as if each connection is tested, to determine its value. The useful circuits are kept; the others are trimmed away, leaving a sleek, efficient machine.

Premature birth likely disrupts these processes—the migration of the nerve cells, the growth of glial cells and white matter, and the trimming. Premature kids have most of the neurons they will carry with them into adult life, but it's possible they're not in the right places or properly connected or tested. Researchers, says Peterson, are "intensively testing" these possibilities.

Peterson's research offers the hope of helping children compensate for whatever brain-related peculiarities they might have. "We want to use imaging to predict who's going to have particularly difficult problems in the course of development, so we can intervene more effectively," he says. That intervention might consist of specially designed education programs or physical therapy and other treatments to compensate for physical and emotional difficulties.

When Peterson began this work, his interest was professional. But now he has a personal interest as well. Two years ago, his daughter was born four weeks premature. While she shows no ill effects, he says he watches her, and he worries.

BRAINSTORMING

When Peterson scanned me, he found nothing wrong or worrisome. If I'd had a brain tumor or some prominent abnormality, he would have spotted it. But that's about all the clinically useful information he could get from a quick scan. If Peterson had put me through the sophisticated scans he uses with the premature infants, perhaps he could have detected some quirk in the way my brain behaves. But because of the large variability in normal brain structure and function, he would not have been able to conclude much specifically about how my brain differs from those of other people.

In the coming years, however, as the technology continues to improve, it may become possible for any of us, with or without obvious illnesses or neurological problems, to learn much more about the state of our brains, our perceptions, and our thinking. "The bad news is that although these techniques are very powerful, they are not where we need to be," says MIT's Desimone. "We need to use these MRI magnets in ways they haven't been used before."

Desimone's McGovern Institute has just inaugurated the Martinos Imaging Center. One room at the center houses a state-of-the-art MRI scanner. Beside it is another room that, for the time being, will remain empty. "We have it sitting there for a new device," Desimone says. He doesn't yet know what that device will be. "That's our challenge—to invent it here. The idea is to go beyond where we are now, to the technology of the future."

The Economics of Brains[*]

A Collection of Research Papers Touts the Promise of Neuroeconomics

By Gregory T. Huang
Technology Review (Cambridge, Mass.), May 2005

Traditional economic theory assumes that human beings behave rationally. That is, that they understand their own preferences, make perfectly consistent choices over time, and try to maximize their own well-being. This peculiar assumption has its roots in dusty essays like "Exposition of a New Theory on the Measurement of Risk" (from 1758) by Daniel Bernoulli and scholarly tomes like Theory of Games and Economic Behavior by John von Neumann and Oskar Morgenstern (published in 1944). The idea has some validity: traditional economic theory is good at predicting some market behaviors, such as how the demand for products like gasoline will change after a tax hike. But it's not very good at describing more-complex phenomena like stock-price fluctuations or why people gamble against the odds.

The problem, of course, is that people don't always behave rationally. They make decisions based on fear, greed, and envy. They buy plasma TVs and luxury vehicles they can't afford. They don't save enough for retirement. They indulge in risky behavior such as gambling. Economists understand this as well as anyone, but in order to keep their mathematical models tractable, they make simplifying assumptions. Then they try to adjust their equations by adding terms that account for "irrational" behavior. But if economists could develop models that accounted for the subtleties of the human brain, they might be able to predict complex behaviors more accurately. This, in turn, might have any number of practical applications: investment bankers could hedge against financial euphoria like the Internet boom; advertisers could sell products more winningly.

The idea that understanding the brain can inform economics is controversial but not new; for 20 years, behavioral economists have argued that psychology should have a greater influence on the development of economic models. What is new is the use of technology: economists, like other researchers, now have at their disposal powerful tools for observing the brain at work. The most popular tool, functional magnetic resonance imaging (fMRI), has been around since the late 1980s; but only in the past few years has it been used to study decision-making, which is the crux of economic theory.

The result is the emerging field of "neuroeconomics." A flurry of recent papers in scientific and economic journals—reviewed in the *Journal of Economic Literature* by Caltech economics professor Colin Camerer and colleagues—shows how researchers are using the neural basis of decision-making to develop new economic models. At the January meeting of the American Economic Association, the world's largest economics conference, the neuroeconomics sessions were reportedly standing room only. The hope seems to be that biological research will finally help economists make sense of irrationality.

Take recent brain-imaging experiments by Princeton University psychologist Samuel McClure. In the journal *Science*, McClure and colleagues report that when subjects choose short-term monetary rewards, different regions of the brain are active than when they choose long-term ones. People don't "discount" future rewards according to a simple scheme, as many economists have suggested. It seems the brain actually makes short-term and long-term forecasts in different ways. The challenge for economists lies in translating this sort of scientific insight into, say, predictive models of how people plan purchases or make retirement fund decisions.

If successful, neuroeconomics could help unify the social sciences and natural sciences—all with great societal impact. "We are at the very beginning of something radically new," says Daniel Kahneman, the Princeton University psychologist who won the 2002 Nobel Prize in economics. "Technologically, we can expect that within the next decade or two there will be huge developments. The network of knowledge about the brain is expanding at a tremendous rate. That will certainly affect marketing and political psychology, and it could create a common database that nobody will want to ignore."

DECISIONS, DECISIONS

It's an intriguing idea: to rethink economic theory from the ground up, taking into account the workings of the human brain. For now, though, neuroeconomics is far removed from the day-to-day concerns of most financiers or CEOs.

The first thing to remember is that the field is very, very young. Neurological tools are still relatively crude. Brain-imaging techniques such as fMRI and positron emission tomography (PET) measure changes in blood flow and hence reveal the collective activity of thousands of neurons over a period of seconds. An electroencephalogram (EEG) uses electrodes on the scalp to measure the brain's electrical activity on the millisecond time scale, but its spatial resolution is so poor that its use is limited. What's more, imaging studies point out only correlations between brain activity and behavior. One must be careful in drawing neuro-scientific conclusions and making economic predictions.

Because their field is so young, and because they are pursuing different goals, economists and neuroscientists working in neuroeconomics sometimes seem to be talking about different things. For instance, Camerer and his colleagues write that "The foundations of economic theory were constructed assuming that details about the functioning of the brain's black box would not be known. . . . [But now] the study of the brain and nervous system is beginning to allow direct measurement of thoughts and feelings." Most neuroscientists would disagree with the second point. Direct measurement of how groups of neurons interact and which brain areas are active during which physical and mental tasks, yes. But thoughts and feelings are subjective (see "The Unobservable Mind," February 2005)and observable only by interpreting data.

In a similar vein, neuroscientists and psychologists have at times equated economic utility—the subjective value of a good or service—with the notions of reward and pleasure. These ideas may be related, but they are certainly not interchangeable. Nevertheless, early mutual confusion about both fields' technical terms and bodies of knowledge is being resolved. "We are rapidly approaching a common language," says Gregory Berns, a neuroscientist at Emory University.

A more fundamental issue for neuroeconomics is this: should economists care? Perhaps understanding how the brain works is more trouble than it's worth. After all, some recent findings are not at first glance very economically enlightening. Anyone who has regretted an impulse purchase, for instance, would be unsurprised to learn that evaluations of immediate and delayed rewards use different parts of the brain. For now, neuroeconomics is subject to the criticisms that plague psychology: that its experiments show what is already intuitively obvious, and its models are descriptive, not quantitative. But Stanford psychologist Brian Knutson and psychiatrist Richard Peterson are trying to answer that criticism. Their paper in a forthcoming issue of *Games and Economic Behavior* reports that subjects seem to use different parts of their brains when they consider financial gains and when they consider financial losses; more recently, they have found that subjects use different parts again to evaluate the magnitude and probability of those gains and losses. Knutson and Peterson's work is part of an increasing effort to figure out how economic utility may be coded quantitatively in various regions of the brain. If economists could track the different components of utility in a statistical way, they could understand why some people take risks and some don't—and possibly predict their future behavior.

PROTECT US FROM OURSELVES

Suppose that the science and technology of neuroeconomics progress according to plan. (They won't, of course, but let's set that aside for now.) At some point in the future, our brains' inner workings, our innermost thoughts, all of our decision-making processes, could be deciphered and displayed individually and unambiguously, like the hands of poker players in televised tournaments. What would we do with this information? How would we protect ourselves? Entire industries—finance, health care, advertising—stand to flourish or die based on the answers.

Let's consider some early indications of what the social consequences of neuroeconomics could be. In finance, an initial attempt at using brain studies to model markets was put forth in a recent paper by the economist Andrew Lo. Lo, the director of MIT's Laboratory for Financial Engineering, argues that the standard theory of "efficient markets"—which assumes investors have perfect information and behave rationally—should be replaced by an "adaptive markets" hypothesis that accounts for psychological factors and responses. He is currently working to formalize the hypothesis mathematically and to implement predictive models of equity risk premium and other stock-market returns using high-performance parallel processors.

Lo is perhaps best known for a study published in 2002 in which he and Dmitry Repin of Boston University used a polygraph-like system to measure the physiological responses of securities traders as they did their jobs; the researchers concluded that emotions like anxiety and fear play a large role in financial decision-making, and that they may have more influence on less experienced workers than on seasoned veterans. "Within five years, neuroeconomics will become mainstream," says Lo. "In 15 to 20 years, it will be fully accepted."

Well before then, expect to see the influence of "neuromarketing" on advertising. Recent experiments have imaged people's brains as they chose between brand names, even movie trailers. Researchers believe that by recording which brain areas are activated during choices, they are starting to be able to predict preferences based on brain scans alone. Some marketing experts believe such research could be used to supplement product surveys and might, eventually, indicate how to ignite pleasurable feelings in consumers at the prospect of rewards.

All of this raises questions about privacy and individual autonomy—and how society might wish to regulate much more effective advertising. "As corporations learn to take further advantage of our weaknesses, we may soon be asking for government to take on the role of protector and guarantor of our privacy, happiness, and savings," says Peterson, who is a managing partner of San Francisco firm Market Psychology Consulting.

That may sound a little excessive. But neuroeconomists are thinking about the influence their work could have on public policy. One of the earliest neuroeconomics papers to address policy implications, "Addiction and Cue-Triggered Decision Processes," by Stanford economists Douglas Bernheim and Antonio Rangel, makes some sensible recommendations. The researchers propose a mathematical theory of addiction (essentially, an economic model) that takes into account findings from brain scans of recovering addicts and physiological measurements from the reward pathways of animal brains. The theory provides a way to determine, for instance, the probability that a recovering alcoholic will drink, depending on the placement of beer cans in a supermarket. It also predicts the effects of addictive-substance policies on the welfare of addicts and casual users—which could be used to compare the socioeconomic consequences of, say, raising taxes on alcohol or subsidizing rehabilitation programs. According to Rangel, this kind of analysis might also apply to other behaviors, like compulsive shopping. The hope is that such models, grounded in the latest neurobiological thinking, will better inform policymakers and lead to more intelligent legislation.

Neuroeconomics seems to be a promising step toward a more unified theory of human behavior. Indeed, by opening up the brain and studying how its circuits produce economic decisions, scientists may provide answers to some of the questions debated by philosophers for centuries. Why do we make the choices we make? And why is it so hard to figure out what we really want?

3

Works in Progress: Brain Development from Infancy to Adulthood

Editor's Introduction

For years, scientists believed that most, if not all, brain development takes place during the first few years of life. By the time children reach adolescence, prevailing theories held, they have long since cemented key neurological connections and charted an unalterable course toward the adults they will eventually become. As the actor and director Rob Reiner put it in a 1997 speech, "[B]y the age of ten, your brain is cooked." Recent findings have led researchers to rethink these assertions, however, and it is now believed that the window of learning remains open throughout an individual's life. While it's true that, by the age of six, our brains are 90 to 95 percent of their adult size, they are "plastic" in nature, constantly changing and rewiring. "I wouldn't disagree with Rob Reiner that the first three years are important," Jay Giedd, the chief of brain imaging in the child-psychiatry division of the National Institute of Mental Health, told Nora Underwood for *The Walrus*. "I would just say that so are the next three and the next three and the next three, up to twenty-five and perhaps even beyond."

These new theories of brain development have far-reaching implications, particularly in the field of education. Whereas some parents still scramble to purchase early-childhood learning tools—toys, books, and videos intended to prepare babies for later academic success—others are taking a more patient approach. Since the brain can't learn new skills, such as reading, until it's ready, scientists contend, it's important to let children develop on their own schedules. Instead of playing Mozart CDs or "Baby Einstein" DVDs, researchers posit, parents should focus on talking to their children, keeping them stimulated, and creating safe and nurturing environments. Studies have shown that children who grow up with neglectful, abusive parents are more likely to develop depression and other disorders. As children progress from infancy to school age, their brains develop and organize neurological connections, keeping those that are used often and "pruning," or discarding, those that are not. This process underscores the importance of exposing young people to a variety of sights and sounds. Teenagers, too, have malleable brains, and studies indicate that, relative to adults, they have difficulty judging risk and realizing the consequences of their actions. Findings also suggest that teenagers' actions are governed more by emotion than reason—something that may come as little surprise to parents of moody adolescents.

In "Shaping the Brains of Tomorrow: What Developmental Science Teaches

About the Importance of Investing Early in Children," the first article in this chapter, Ross A. Thompson explains why it's critical for parents to create positive experiences for their children. Children learn language skills and basic morals by observing and interacting with others, and those that grow up in unstable environments often have trouble adjusting to social situations. In his piece "How the Arts Develop the Young Brain," the second selection in this chapter, David Sousa considers how exposure to art promotes learning in a variety of academic subjects. Sousa cites research that suggests young people who learn to play musical instruments fare better on reading and mathematics tests.

The chapter continues with "The Teenage Brain," in which Nora Underwood challenges the notion that our brains are ever truly "cooked." Using the results of recent MRI experiments, she explores the many ways in which teenagers differ from adults. In the final article, "When Does Your Brain Stop Making New Neurons?" Sharon Begley examines how brains continue to evolve into adulthood and old age. In discrediting "neurological nihilism"—the idea that, after a certain age, our intelligence and character traits are set in stone—Begley holds out hope that people can better themselves in ways previously thought impossible.

Shaping the Brains of Tomorrow[*]

What Developmental Science Teaches About the Importance of Investing Early in Children

By Ross A. Thompson
The American Prospect, November 2004

What would happen if the best minds in the country concluded that investments in early-childhood development are necessary and cost-effective? That the early years present an opportunity, unequaled later in life, to enhance inborn potential and avert harm? What if they could identify the "active ingredients" of healthy psychological development, and how to enhance these in young children growing up in deprived conditions? Wouldn't society become mobilized to do its best for young children?

We are in this situation today, and the arguments for investing in early-childhood development are scientific, not political. As the result of several blue-ribbon studies of the forces shaping young children's growth, developmental scientists today agree on some basic conclusions: The early years are important. Early relationships matter. All children are born ready to learn, both intellectually and socially. Even in infancy, children are active participants in their own development, together with the adults who care for them. Early experience can elucidate, or diminish, inborn potential. The early years are a period of considerable opportunity for growth and vulnerability to harm.

What we do with this knowledge will shape the lives of the next generation.

DEVELOPMENT IN THE EARLY YEARS

Developmental psychologists and neurobiologists agree that the developing mind is astonishingly active and self-organizing, creating new knowledge from everyday experiences. Newborns crave novelty and become bored with familiarity, so their eyes, ears, and other sensory organs are attuned to events from which they can learn. A few months later, the infant mentally clusters objects together that are similar in shape, texture, or density, and explores gravity and causality as crackers are dropped from the high chair. A toddler categorizes faces, animals, and birds according to their properties, and by age 3 or 4, children make logical inferences about new members of a group, such as appreciating that a dolphin breathes like the mammal it is rather than the fish it resembles. Just as the developing brain is expanding its interconnections, the developing mind is making connections between the new knowledge it discovers and creates.

The remarkable intellectual accomplishments of the early years extend to language development. Newborns have an innate capacity to differentiate speech sounds that are used in all the world's languages, even those they have never heard and which their parents cannot discriminate. But later in the first year they lose this ability as they become perceptually attuned to the language they will learn. By age 3, a child is forming simple sentences, mastering grammar, and experiencing a "vocabulary explosion" that will result, by age 6, in a lexicon of more than 10,000 words. Equally important, language will enable the child to put developing ideas and concepts into words that he or she can share with another, revolutionizing his or her thought by gaining access to the concepts, ideas, and values of others.

Sensitive caregiving—not educational toys or Mozart CDs—provides the most essential catalysts for these feats of intellectual growth. People are critical to the development of the mind: Newborns attend in a special way to human faces and voices, toddlers learn new words based on their interest in the intentions of adult speakers, and memory develops through the shared recounting of everyday events. Relationships stimulate the mind and provide the emotional incentives to new learning as young children share their discoveries with another. This is why promoting school readiness is not simply a matter of encouraging literacy and number skills. It must also ensure the secure, unhurried, focused attention from sensitive caregivers that contributes to the growth of curiosity, the eagerness to discover, self-confidence, and cooperation.

Healthy brain development relies on people to provide the stimulation that organizes connections in the cortex for language and complex thought. It also relies on people to protect the baby from overwhelming stress, manage the child's emotions, and promote security. This is why strong attachments between infants and their caregivers are as biologically basic as learning to crawl and walk. Throughout evolution, attachment relationships have ensured human survival by keeping infants protected and nurtured. By their first birthday, infants have developed deep attachments to those who care for them. And these attachments, in turn, provide a foundation for positive relationships with peers and teachers, healthy self-concept, and emotional and moral understanding.

In the absence of nurturing relationships, things can go wrong. It isn't surprising to find that insecure attachments develop more frequently in homes where parents are stressed or depressed, or in chaotic child-care settings. Even more disturbing is research demonstrating how early children show signs of depression, conduct problems, social withdrawal, and anxiety disorders, and how closely these problems are tied to the quality of the parent-child relationship. These studies show that relationships with caregivers who are neglectful, physically abusive, or emotionally troubled can predispose young children to psychopathology. So the importance of these earliest relationships is a double-edged sword: Sensitive caregiving underpins healthy development, while markedly inadequate care renders young children vulnerable to harm.

Relationships also influence the growth of social and emotional understanding. Far from being egocentric, young children are fascinated by what goes on in others' minds, and social experiences are the laboratory in which these discoveries emerge. A 2-year-old whose hand inches closer to the forbidden VCR while carefully watching her parent's face, for example, is testing the adult's expected reaction. And a 3-year-old whose roughhousing has resulted in a crying younger sibling learns from an adult about the connections between exuberant running and inadvertent collisions, enhancing his or her emotional understanding and empathy.

FROM MIND TO BRAIN—AND BACK AGAIN

Whether we are concerned with the growth of the mind or the person, all of these remarkable early achievements take place in the developing brain. Brain development begins within the first month after conception, and by the sixth prenatal month, nearly all of the billions of neurons that populate the mature brain have been created. This means that the quality of prenatal care, particularly the mother's nutrition, health, and exposure to dangerous viruses and drugs, can have a profound effect on the developing brain of her fetus. Health, nutrition, and drug exposure continue to influence brain development after birth.

Both before and after birth, there is an initial "blooming" of connections between neurons, creating a brain densely packed with many more neural pathways than it needs. This proliferation is followed by a period of "pruning" in which little-used connections gradually erode to reach the number required for optimal efficiency. Experience is the central determinant of which neural pathways are retained or disappear. The early experiences that sculpt the developing brain can be stimulating or neglectful, supportive or traumatic, secure or stressful. Through a "use it or lose it" principle, those neurons that aren't activated through experience progressively wither. Language exposure, for example, helps to account for the transition from the newborn's capacity to perceive universal speech sounds to the 1-year-old's language-specific speech perception. Developmental neuroscientists offer similar accounts to explain the early development of vision, memory ability, early categorization and thinking skills, and emotional development.

Brain development is an extended process—not limited to a narrow "window of opportunity" between zero and three, as conventional wisdom sometimes suggests. Neural connections in areas of the brain guiding higher forms of thinking and reasoning grow and atrophy into early adolescence, for example, and the adult brain even creates new neurons in certain regions governing memory. Brain architecture continues to be subtly refined throughout life in ways that reflect the individualized, everyday experiences of the person. The brain of a musician who plays a stringed instrument, for example, differs from the brain of a poet who works with words and abstract ideas because they have exercised different brain regions throughout life.

Despite these exciting discoveries, neuroscientists are still at the early stages of understanding how experience refines the brain. They are concerned with how early deprivation (such as that experienced by orphans from Romania and the former Soviet Union), abuse, and trauma influence early brain growth, and whether these effects can be altered. They are also studying how relational problems, such as the challenges faced by an infant of a depressed mother, influence brain development.

INVESTING IN YOUNG CHILDREN

These and other conclusions from a landmark study of the National Academy of Sciences, From Neurons to Neighborhoods: The Science of Early Childhood Development, underscore the importance of early experiences for development throughout life.

What about children, then, who live in deprived or high-risk conditions? Considerable research shows that many of them will lag intellectually from infancy and will suffer deficiencies in various facets of healthy psychological development. Poverty significantly compromises healthy intellectual and socioemotional development, for example, and poverty during early childhood is more powerfully predictive of later achievement than is poverty at any later stage. The reasons include stressed caregivers, troubled parent-child relationships, dangerous neighborhoods, and inadequate schools and community supports.

Can early interventions improve the odds of healthy development for children at risk? The answer offered by the committee of scientists that wrote From Neurons to Neighborhoods is both optimistic and challenging. The good news is that there are successful strategies, especially programs that emphasize child-focused educational activities and parent-child interaction, and are governed by specific practices matched to clear goals. But the most effective interventions are rarely simple, inexpensive, or easy to implement. Changing the developmental trajectory of a young child growing up in deprived circumstances requires determination, persistence, and patience.

Are such interventions cost-effective? Determining the cost-effectiveness of programs for at-risk young children requires putting price tags on the innumerable human consequences of early deprivation. Yet several studies of comprehensive early-intervention efforts have found that program costs are more than compensated by averted costs of educational remediation, juvenile or adult crime, and diminished job earnings.

While expensive, large-scale public efforts have been skeptically regarded by policy-makers most concerned about their costs, important new voices are emerging in support of these investments. One is that of James Heckman, Nobel laureate and University of Chicago economist, who argues that the varied benefits of early-childhood interventions—in cognitive learning, motivation, and socialization—are likely to have long-term advantages in the labor market because of the cumulative effects of early improvements in ability. Another is that of Art Rolnick of the Federal Reserve Bank in Minneapolis, who (along with colleague Rob Grunewald) estimates that public investments in programs to assist poor children yield a 16-percent real rate of return. This, he argues, compares very favorably to other public investments with more popular appeal, such as building sports coliseums, which typically have little or no return on public investment. Although much more research is needed, it appears that society's investment in improving the chances for young children at risk is economically worthwhile.

The views of economists like these shift the debate about public efforts to support healthy early development. And they join the chorus of scientists whose work has consistently shown how much early-childhood experiences and relationships matter. It is now reasonable to ask why public policy lags so significantly behind the science and economics of early-childhood development.

The public policies that would support healthy early-childhood development are child-friendly and family-friendly. They include:
- child-care policies that ensure widespread access to affordable, high-quality child care;
- welfare-reform policies that enable parents to integrate work and family responsibilities constructively in children's interests;
- prenatal and postnatal health care that screens children for developmental difficulties before they become severe, guarantees adequate nutrition, provides early visual and auditory screening, and protects young children from debilitating diseases and hazardous exposure to environmental toxins.

In the end, because children are society's most valuable asset, they are also a social responsibility and investment. Because the science of early-childhood development converges with the economics of public policy to confirm that investments in early-childhood development are both necessary and worthwhile, it is long past time for society to catch up.

How the Arts Develop the Young Brain[*]

By David Sousa
The School Administrator, December 2006

Every culture on this planet has art forms. Why is that? Neuroscientists continue to find clues as to how the mental and physical activities required for the arts are so fundamental to brain function.

Certain brain areas respond only to music while others are devoted to initiating and coordinating movement from intense running to the delicate sway of the arms. Drama provokes specialized networks that focus on spoken language and stimulate emotions. Visual arts excite the internal visual processing system to recall reality or create fantasy with the same ease.

These cerebral talents are the result of many centuries of interaction between humans and their environment, and the continued existence of these talents must indicate they contribute in some way to our survival. In those cultures without reading and writing, the arts are the media through which that culture's history, mores and values are transmitted to the younger generations and perpetuated. They also transmit more basic information necessary for the culture's survival, such as how and what to hunt for food and how to defend the village from predators. Here, art becomes an important force behind group survival.

In modern cultures, the arts are rarely thought of as survival skills, but rather as frills—the esthetic product of a wealthy society with lots of time to spare. People pay high ticket prices to see the arts performed professionally, leading to the belief that the arts are highly valued. This cultural support often is seen in high schools, which have their choruses, bands, drama classes and an occasional dance troupe.

Yet seldom do public elementary schools enjoy this continuous support. When school budgets get tight, elementary-level art and music programs are among the first to be reduced or eliminated. Now, pressure from the No Child Left Behind Act to improve reading and mathematics achievement is prompting elementary schools to trade off instruction in the arts for more classroom preparation for the mandatory high-stakes tests. Ironically, this is happening just when neuroscience research is revealing the impressive impact that the arts have on the young brain's cognitive, social and emotional development.

COGNITIVE GROWTH

During the brain's early years, neural connections are being made at a rapid rate. Much of what young children do as play—singing, drawing, dancing—are natural forms of art. These activities engage all the senses and wire the brain for successful learning.

When children enter school, these art activities need to be continued and enhanced. Brain areas are developed as the child learns songs and rhymes and creates drawings and finger paintings. The dancing and movements during play develop gross motor skills, and the sum of these activities enhances emotional well-being. And sharing their artwork enhances social skills.

The arts are not just expressive and affective, they are deeply cognitive. They develop essential thinking tools—pattern recognition and development; mental representations of what is observed or imagined; symbolic, allegorical and metaphorical representations; careful observation of the world; and abstraction from complexity.

The arts also contribute to the education of young children by helping them realize the breadth of human experience, see the different ways humans express sentiments and convey meaning, and develop subtle and complex forms of thinking. Although the arts are often thought of as separate subjects, like chemistry or algebra, they really are a collection of skills and thought processes that transcend all areas of human engagement.

MUSIC LISTENING

Many researchers believe the ability to perceive and enjoy music is an inborn human trait. This biological aspect is supported by the discovery that the brain has specialized areas that respond only to music and that these areas provoke emotional responses. Brain scans show the neural areas stimulated depend on the type of music—melodic tunes stimulate areas that evoke pleasant feelings while dissonant sounds activate other areas that produce unpleasant emotions.

Research studies show that before infants reach their first birthday, they are able to use music as a retrieval cue, differentiate between two adjacent musical tones, recognize a melody when it is played in a different key and categorize rhythmic and melodic patterns on the basis of underlying tempo. Research on the effects of music on the brain and body are divided into the effects of listening to music and the effects of creating or producing music on an instrument.

The notion that music affects cognitive performance catapulted from the research laboratory to the television talk shows in 1993 when a study found that spatial-temporal reasoning—the ability to form mental images from physical objects or to see patterns in time and space—improved in college students after listening to a Mozart sonata for 10 minutes. However, the media failed to report that the students' improved abilities faded within an hour. The results of this study, promptly dubbed "The Mozart Effect," were widely publicized and misinterpreted to imply that listening to a Mozart sonata would enhance intelligence by raising IQ.

Subsequent studies have confirmed that listening to Mozart does enhance various types of spatial and temporal reasoning tasks, especially problems requiring a sequence of mental images to correctly reassemble objects. The data suggest that the effect is real, yet it occurs with other kinds of music besides Mozart. Researchers, however, do not yet know conclusively why the effect occurs. Nonetheless, the effect is important to educators because it shows that passive listening to music stimulates spatial thinking and that neural networks normally associated with one kind of mental activity readily share the cognitive processes involved in a different activity. In other words, learning or thinking in one discipline may not be completely independent of another.

Other studies have shown that listening to certain music stimulates the parts of the brain responsible for memory recall and visual imagery. Researchers have also found that listening to background music enhances the efficiency of those working with their hands. This explains why background music in the classroom helps many students stay focused while completing specific learning tasks. Overly stimulating music, however, serves more as a distraction and interferes with cognitive performance.

CREATING MUSIC

Although passive listening to music has short-term educational benefits, creating instrumental music provides many more cerebral advantages. Learning to play a musical instrument challenges the brain in new ways. In addition to being able to discern different tone patterns and groupings, new motor skills must be learned and coordinated in order to play the instrument correctly. These new learnings cause profound and seemingly permanent changes in the brain, and certain cerebral structures are larger in musicians than in non-musicians.

This raises an intriguing question: Are the brains of musicians different because of their training in music, or did these differences exist before they learned music? The answer came when researchers trained non-musicians to listen for small changes in pitch and similar musical components. In just three weeks, their brains showed increased activation in the auditory processing regions. This suggests the brain differences in highly skilled musicians are more likely the result of training and not inherited. No doubt some genetic traits enhance music learning, but it seems most musicians are made, not born.

The beneficial effects of learning to play an instrument begin at an early age.

One major study involved 78 children from three California preschools, including one school serving mostly poor, inner-city families. The children were divided into four groups. One group took individual, 12- to 15-minute piano lessons twice a week. Another group took 30-minute singing lessons five days a week, and a third group trained on computers. The fourth group served as the control and received no special instruction. All students took tests before the lessons began to measure different types of spatial-reasoning skills.

After six months, the children who received piano keyboard training had improved their scores by 34 percent on tests measuring spatial-temporal reasoning. On other tasks, there was no difference in scores. Furthermore, the enhancement lasted for days, indicating a substantial change in spatial-temporal function. The other three groups, in comparison, had only slight improvement on all tasks. Subsequent studies continue to show a strong relationship between creating music with keyboards and the enhancement of spatial reasoning in young children.

In addition, numerous studies have shown that musical training improves verbal memory. Researchers in one study administered memory tests to 90 boys between the ages of 6 and 15. Half belonged to their school's strings program for one to five years, while the other half had no musical training. The musically trained students had better verbal memory. Furthermore, the memory benefits of musical training were long-lasting. Students who dropped out of the music training group were tested a year later and found to retain the verbal memory advantage they had gained earlier.

BETTER NUMERACY

Of all academic subjects, mathematics is most closely connected to music. Counting is fundamental to music because one must count beats, count rests and count how long to hold notes. Music students use geometry to remember the correct finger positions for notes or chords on instruments. Reading music requires an understanding of ratios and proportions so that whole notes are held longer than half notes.

Music and mathematics also are related through sequences called intervals: A mathematical interval is the difference between two numbers and a musical interval is the ratio of their frequencies. And arithmetic progressions in music correspond to geometric progressions in mathematics.

Several imaging studies have shown that musical training activated the same areas of the brain that were also activated during mathematical processing. It appears that early musical training begins to build the same neural networks that later will be used to complete numerical and mathematical tasks.

To further study this idea, researchers sought to determine whether learning to play a piano keyboard would help young students learn specific mathematics skills. They focused on proportional mathematics, which is particularly difficult for many elementary students and which is usually taught with ratios, fractions and comparative ratios. One group of 2nd-grade students from a low socioeconomic Los Angeles neighborhood was given four months of piano keyboard training along with computer training on software designed to teach proportional mathematics. This group scored 166 percent higher on proportional mathematics and fractions subtests than the matched group that received neither music nor specific computer lessons, but did play with the computer software. These findings are significant because proportional mathematics is not usually introduced until 5th or 6th grade and because a grasp of proportional mathematics is essential to understanding science and mathematics at higher grade levels.

Another study found that low socioeconomic students in California who took music lessons from 8th through 12th grade increased their test scores in mathematics and scored significantly higher than those low socioeconomic students who were not involved in music. Mathematics scores more than doubled, and history and geography scores increased by 40 percent.

A subsequent review of studies involving more that 300,000 secondary school students confirmed the strong association between music instruction and achievement in mathematics. Of particular significance is an analysis of six experimental studies that revealed a causal relationship between music and mathematics performance and that the relationship had grown stronger in recent years.

Educators might want to consider this relationship in planning the core curriculum. If numeracy is so important, perhaps every student should learn to play a musical instrument.

READING CONNECTIONS

Several studies confirm a strong association between music instruction and standardized tests of reading ability. Studies conducted with 4- and 5-year-old children revealed that the more music skills children had, the greater their degree of phonological awareness and reading development.

Apparently, music perception taps and enhances auditory areas that are related to reading. Although we cannot say that taking music instruction caused the improvement in reading ability, this consistent finding in a large group of studies builds confidence that there is a strong relationship. Researchers suggest this relationship results because both music and written language involve similar decoding and comprehension reading processes and require a sensitivity to phonological and tonal distinctions.

In the area of the visual arts, the human brain has the incredible ability to form images and representations of the real world or sheer fantasy within its mind's eye. Solving the mystery of DNA's structure, for example, required James Watson and Francis Crick in the early 1950s to imagine numerous three-dimensional models until they hit on the only image that explained the molecule's peculiar behavior— the spiral helix. This was an incredible marriage of visual art and biology that changed the scientific world forever.

Exactly how the brain performs the functions of imagination and meditation may be uncertain, but no one doubts the importance of these valuable talents, which have allowed human beings to develop advanced and sophisticated cultures.

IMAGE PRODUCERS

Although teachers spend much time talking about the learning objective, little time is given to developing visual cues. This process, called imagery, is the mental visualization of objects, events and arrays related to the new learning and represents a major method for storing information in the brain.

Imagery takes place in two ways: imaging is the visualization in the mind's eye of something that the person has actually experienced; imagining depicts something the person has not yet experienced and, therefore, has no limits. The research evidence is clear: Individuals can be taught to search their minds for images and be guided through the process to select appropriate images that enhance learning and increase retention.

As students today engage with electronic media that produce external images, they are not getting adequate practice in generating their own internal imaging and imagining, skills that not only affect survival but also increase retention and, through creativity, improve the quality of life.

Imagery can be used in many classroom activities, including visualized notetaking, cooperative learning groups and alternative assessment options. Mindmapping is a specialized form of imagery that combines language with images to show relationships between and among concepts and how they connect to a key idea.

Coaches have known for a long time that athletes who use imagery to mentally rehearse what they intend to do perform better than if they do not use imagery. Studies reveal that the more time and intensity devoted to imagery, the better the athletic performance.

Apart from sports, data from nine studies involving nearly 1,500 students were analyzed and showed a statistically significant association between imagery and creativity. Not surprisingly, students who used more imagery during learning displayed more creativity in their discussions, modeling and assessments.

PHYSICAL ACTIVITY

Even short, moderate physical exercise improves brain performance. Studies indicate that regular physical activity increases the number of capillaries in the brain, thus facilitating blood transport. It also increases the amount of oxygen in the blood, which significantly enhances cognitive performance. Despite the realization that physical activity enhances brain function and learning, secondary students spend most of their classroom time sitting. Although enrollment in high school daily physical education classes has risen slightly in recent years, it represents only about 25 percent of the student body.

Teachers need to encourage more movement in all classrooms at all grade levels. At some point in every lesson, students should be moving about, talking about their new learning. Not only does the movement increase cognitive function, but it uses up some kinesthetic energy so students can settle down and concentrate better later. Mild exercise before a test also makes sense. So does teaching dance to all students in K–8 classrooms. Dance techniques help students become more aware of their physical presence, spatial relationships, breathing, and of timing and rhythm in movement.

Movement activities are also effective because they involve more sensory input, hold the students' attention for longer periods of time, help them make connections between new and past learnings and improve long-term recall. We can easily recall the time we participated in the school play or other public performance because this experience activated our kinesthetic sensory system. Moreover, many students are involved with interesting kinesthetic activities after school. Doing these types of activities in school awakens and maintains that interest.

ARTS INTEGRATION

Research studies have examined both stand-alone arts programs as well as those that integrate concepts and skills from the arts into other curriculum areas. One intriguing revelation of these studies is that the most powerful effects are found in programs that integrate the arts with subjects in the core curriculum. Researchers suggest that arts integration causes both students and teachers to rethink how they view the arts and generates conditions that are ideal for learning.

Studies consistently show the following in schools where arts are integrated into the core curriculum: Students have a greater emotional investment in their classes; students work more diligently and learn from each other; cooperative learning groups turn classrooms into learning communities; parents become more involved; teachers collaborate more; art and music teachers become the center of multi-class projects; learning in all subjects becomes attainable through the arts; curriculum becomes more authentic, hands-on and project-based; assessment is more thoughtful and varied; and teachers' expectations for their students rise.

The arts play an important role in human development, enhancing the growth of cognitive, emotional, and psychomotor pathways. Schools have an obligation to expose children to the arts at the earliest possible time and to consider the arts as fundamental (not optional) curriculum areas. Finally, learning the arts provides a higher quality of human experience throughout a person's lifetime.

The Teenage Brain[*]

Why Adolescents Sleep in, Take Risks, and Won't Listen to Reason

By Nora Underwood
The Walrus, November 2006

You don't have to suffer to suffer to be a poet. Adolescence is enough suffering for anyone.
—American poet John Ciardi, 1962

In his speech at the launch of the 1997 I Am Your Child campaign, director and actor Rob Reiner stated that "by the age of ten, your brain is cooked." And until recently, most child experts, including Dr. Spock, would have agreed. They considered the first few years of a child's life to be the most important—and the experiences a child had during those years to play a crucial role in defining the kind of person he or she would ultimately become. That understanding also helped create a whole generation of obsessively child-focused parents, who, with the best of intentions, have tried to cram a lifetime of "educating" into a few short years, subjecting their unwitting fetuses to a diet of *Eine kleine Nachtmusik* and their pre-verbal toddlers to basic arithmetic and multiple viewings of Baby Einstein DVDs. (A wise elementary-school principal once noted, "I very much doubt Einstein was doing any of this when he was young.")

Somewhere along the line, or so many of us believed, the window of opportunity would close. The foundations of the adult-to-be would be laid, and the worst damage would be done. The majority of brain development does, in fact, take place in the early years, when billions of synaptic circuits that will last the child's lifetime are forming. But growth and change don't end there. Important developmental changes, scientists are discovering, are still taking place in a big way through the adolescent years—and into the mid-twenties. Perhaps this helps to explain the growing phenomenon of adult children who linger on under the parental roof; their growing may not be over, despite their arrival at "adulthood."

In recent years, researchers have finally been able to get real insight into the workings of the brain thanks to magnetic resonance imaging (MRI), using the technology to map blood flow to the areas of the brain that are activated by exposure to various stimuli. By scanning the same group of adolescents over a period of years or by comparing the brain responses of teenagers to those of adults, researchers are putting together a portrait of adolescence that confirms what many parents have always suspected: adolescents might as well be a whole different species. They are, as one neuroscientist puts it, a "work-in-progress."

Over the past decade, scientists have started to grasp exactly how distinctive the adolescent brain is and how crucial the years between ten and twenty-five are in terms of its development. And their discoveries have implications not only for parents, educators, and the medical community but also for policymakers. "I wouldn't disagree with Rob Reiner that the first three years are important," says Jay Giedd, chief of brain imaging in the child psychiatry branch of the National Institute of Mental Health in Bethesda, Maryland. "I would just say that so are the next three and the next three and the next three, up to twenty-five and perhaps even beyond."

This news may not come as a surprise to the mother who still lies awake at 3 a.m., waiting for her basement-dwelling, twenty-two-year-old post-grad son to come home. What science suggests is that "adulthood" as we have defined it doesn't necessarily signal the end of childhood development—or of parental worries.

If the media is to be believed, the stereotypical teen is a selfish, volatile, rude, rebellious hormone-head, capable of little more than taking outrageous risks, ingesting too many harmful substances (legal and otherwise), committing crimes, crashing parties, trashing houses, and generally being a layabout. Of course, this is a gross misrepresentation: many teenagers pass through adolescence smoothly and happily, without becoming parents themselves, dropping out of school, or acquiring a criminal record instead of a degree. Still, there's a stubborn tendency in the culture to ascribe every negative teen moment to "hormones." Recent brain research, however, relieves hormones of much of the blame for this period of "storm and stress," as psychologist G. Stanley Hall, father of adolescent research, called it.

The full extent to which hormones actually influence adolescent behaviour remains unknown. So is what role they play in brain development. Hormones are certainly responsible for the most obvious hallmarks of puberty; at some mysterious point in a child's life, a protein called kisspeptin causes the hypothalamus—an area in the brain that orchestrates certain autonomic nervous-system functions—to secrete the gonadotropin-releasing hormone, which sets the pubertal changes in motion. Ultimately, estrogen and testosterone are responsible for the physical transformations—breast and genital development, body-hair growth, deepening of the voice, and so on—but by no means all the behavioural changes of adolescence. Hormones may have nothing to do with the fact that your daughter can't bear your singing voice, for instance; it's a safe bet, however, that a teenager's fixation on sex and social standing is pretty much hormone related.

But puberty does have an impact on how they think. For instance, as Giedd points out, boys fairly predictably base their decisions on the question "Will this lead to sex? " Giedd adds: "They may not say it in that way or it may not be that blatant, but if you just sort of go with that model it works pretty well." When girls make decisions, he adds, they are more likely to keep the social group, and their place in it, in mind. But Giedd feels that puberty's influence doesn't extend much outside that realm. "Your ability to do a logic problem or to do geometry or to do other things seems to be more [related to] age itself." Researchers have also found that the onslaught of testosterone in both male and female adolescents at puberty literally swells the amygdala—the brain centre associated with the emotions. Perhaps we can blame the amygdala for the slammed doors and sudden tears that overcome previously sunny children when they hit adolescence.

So hormones are not the only players in the changes that characterize adolescence. And while it is difficult to tease out the varying roles played by chromosomes, hormones, and other factors in teen behaviour, the insights that MRI reveals are nothing short of astounding.

Jay Giedd has been using MRI since 1991 to understand how the brain develops from childhood through adolescence and into early adulthood. He has scanned the brains of about 1,800 children, teenagers, and young adults every two years and interviewed them about their lives and feelings. As it turns out, Dr. Spock was not entirely wrong: by the time a child reaches the age of six, the brain is 90 to 95 percent of its adult size. But massive changes continue to take place for at least another fifteen years. They involve not just the familiar "grey matter," but a substance known as "white matter," the nerve tissue through which brain cells communicate—literally the medium that delivers the messages. White matter develops continuously from birth onward, with a slight increase during puberty. In contrast, grey matter—the part of the brain responsible for processing information, or the "thinking" part—develops quickly during childhood and slows in adolescence, with the frontal and temporal lobes the last to mature.

And this is the crux: the frontal lobe, or more precisely the prefrontal cortex, is the home of the so-called "executive functions": planning, organization, judgment, impulse control, and reasoning. The part that should be telling the sixteen-year-old not to dive off the thirty-foot cliff into unknown water. The seat of civilization.

What Giedd has witnessed via MRI is a constant push and pull in the grey matter. Certain forces cause a process known as arborization, during which grey matter gets bushier and grows new dendrites. Balancing that is a regressive pull, a competition for survival of sorts, in which some branches of the grey matter thrive while others are sacrificed. Both processes are continuous; as some new pathways grow, others are being pruned back. The quantity of grey matter peaks in girls around the age of eleven and in boys around thirteen, after which the amount of white matter increases. As grey matter decreases, there is also an increase in myelination, a process during which neurons, or nerve fibres, are insulated to enhance their performance.

In the end, though, the amount of grey matter isn't really the issue. "It's much more related to quality than quantity," explains Giedd. "This pruning process is normal and natural and healthy in terms of optimizing the brain for different environments. Our brains are built to be very adaptable during the teen years"—just the time when children start to figure out how to make it in the world. "The brain is incredibly plastic, which allows us to make it at the North Pole or the equator, to use a computer versus hunting with a stick. The teen brain is able to make changes depending on the demands of the environment." (This might explain a thirteen-year-old's ability to easily master new technology while parents struggle with the TV remote.)

What determines the fate of a cell is whether it has made a meaningful connection with other cells. This is a real use-it-or-lose-it process. As some scientists have noted, if an adolescent forgoes reading in favour of lying around on the couch playing video games, those unused synapses will be pruned. Nobel Prize-winning scientist Gerald Edelman has called this "neural Darwinism"—the survival of the fittest synapses. So scientists know that different activities—playing sports, speaking a second language, drinking, smoking, and so on—influence how the adolescent's brain will ultimately be wired, though they aren't clear what the implications are: Is the pianist going to do better in life than the crossword-puzzle fiend? Will the jock have a leg up, brain-wise, on the geek? "Can you actually see changes in the brain of someone doing music? The answer to that is yes," says Giedd. "But is that a good thing particularly? Is it just that our brains will become specialized in whatever we spend our time doing or is there a more general benefit?"

A father compliments his thirteen-year-old daughter on her new dress, only to have her swivel around, glare at him, and hiss, "What's that supposed to mean?" Nervous parents can rarely tell when an adolescent is going to fly off the handle. Why do they often have such hair-trigger responses? Two different MRI studies indicate that teenagers do not process emotion the same way adults do. In fact, one study shows that the adolescent brain actually reads emotion through a different area of the brain. Dr. Deborah Yurgelun-Todd, director of neuropsychology and cognitive neuroimaging at McLean Hospital in Belmont, Massachusetts, has scanned both adults and teenagers as they were shown images of faces that are clearly expressing fear. All the adults correctly identified the emotion; many of the teens got it wrong (about half labelled the expression one of "shock," "sadness," or "confusion"). Yurgelun-Todd found that during the scan of the adults, both the limbic area of the brain—the area especially connected to emotions—and the prefrontal cortex lit up. When teens were seeing the same pictures, the limbic area was bright but there was almost no activity in the prefrontal cortex. They were having an emotional response essentially unmediated by judgment and reasoning.

In another brain-imaging study, Daniel Pine, a researcher at the National Institute of Mental Health, tried to determine how the brain was able to stay focused on a task while the subject was being exposed to faces that were registering strong emotion. The result: activity in the frontal cortex of the adults was steadier, indicating they were better able to stay on task than teenagers. The emotional faces seemed to activate key areas in the brains of both age groups but only the adults were able to mute that activity so they could stay focused. Teenagers are more at the mercy of their feelings.

There is another fascinating phenomenon that plays havoc with the family of a teen: the adolescent sleep pattern. Suddenly, the kid who always woke you up at sunrise, when you were desperate to sleep, turns thirteen or fourteen and can neither be dragged from bed in the morning nor forced into it at night. Making matters worse, this change invariably occurs as the sleep needs of the middle-aged parents are flipping around the other way. It may seem like just another case of teenage passive aggression, but it's just biology; the circadian rhythm of the brain has changed and teenagers simply don't want to—or can't—go to bed before 12 or 1 a.m.

Why this happens has been the focus of some interest. Researchers at Brown University and Bradley Hospital in Providence, Rhode Island, measured the amount of melatonin, the hormone that helps regulate the sleep-wake cycle, in teenagers' saliva over the course of the day. They discovered that the levels of the hormone increased later in the day and decreased later in the morning in teenagers than in adults and children. A separate study indicated that the biological trigger for sleep—called the sleep pressure rate—slowed down during adolescence.

So if teenagers appear to be cycling through the day at a different pace from the rest of the world, it's because they are. In fact, because they are waking up when the world dictates—rather than when their bodies tell them to—teenagers are chronically sleep-deprived, which can have consequences ranging from superficial to severe. For starters, as Carlyle Smith, a psychology professor at Trent University in Peterborough, Ontario, who has studied how the adolescent brain processes information during sleep, notes, "They're just sleepy." They go to school tired, unfocused, and—because nobody likes to eat breakfast when they'd rather be sleeping—typically unfed. And as many teachers can attest, teenagers are also generally less able to absorb information in the morning. But by later in the afternoon, as the rest of the world is struggling not to nod off at their desks, teenagers begin to fire on all cylinders. "[As an adult] your temperature is at its high point shortly after lunch," explains Smith, "and then it starts its way down and drops all night until 3 or 4 a.m., when it starts to go up again. Theirs doesn't reach its height until later in the day." As a result, teenagers are just starting to focus and become more verbally adept as the rest of the world is crashing. By midnight, while the rest of the family is doing its best to fall asleep, teenagers are wide awake and instant-messaging away.

What is the fallout from a world that runs against the adolescent clock? There are four non-REM stages of sleep, and stages three and four, the deepest, which occur during the first third to first half of the night, are particularly useful to adolescents, who still have those frontal lobes to myelinate and lots of overall growing left to do (growth hormone is released during deep sleep). But because teenagers are so often deprived of REM sleep, which occurs during the last part of the night, their memories can suffer; they lose out on the stage of sleep that sees the information they've absorbed throughout the day replayed and consolidated.

"Kids should be getting over nine hours of sleep," says Smith. "Most are getting one to two hours less than they should. They're missing quite a chunk of REM sleep and that's important for understanding new things. If you don't get much REM sleep, you're not going to learn as fast as people who do."

In one study, Smith set his subjects, who ranged in age from eighteen to twenty-two, to learning a logic task and then deprived them of the last half of the night of sleep. A week later, after the participants had recovered, the researchers tested them again. All had forgotten between 20 and 30 percent of what they'd learned. Once in a while, this kind of sleep loss is no problem. People can catch up. But when sleep deprivation becomes chronic, the consequences are compounded. "You're forgetting 20 percent, but 20 percent every day," says Smith. "And that goes on for months and months and months. That's an inefficient system."

Chronic sleep deprivation also increases the risk of developing depression (though, paradoxically, if someone is already depressed, sleep deprivation tends to help them feel better). This is a particularly serious issue for adolescents, as certain mental-health disorders tend to manifest themselves during these years. "There's so much confusion over this," Smith admits, "but one of the worries is if you just keep on with the sleep deprivation, eventually [that person] will become depressed. And we're seeing a lot more depressed kids around now."

But it's not easy to fight nature; perhaps the best parents can do is to encourage a slowdown of activity at a reasonable time in the evening, keep technology out of the bedroom and caffeine out of the fridge, and let their kids catch up on weekends.

Most adults know what they're up against because they remember their own night-owl days. They may have dabbled in rule-breaking, underage drinking, and general wildness as teenagers and now they shudder at the thought of their own children doing the same or worse. They were lucky, but will their kids inherit their luck? (The bad news for former hellraisers: some research suggests a person's tendency to take risks is partly genetic.)

In fact, there's some indication that cultivating unhealthy habits through this whole tumultuous period of development can have serious long-term effects. Those who start smoking during adolescence, for example, will likely have a much harder time quitting later in life than those who take up smoking in their twenties; the addiction, according to researchers at Duke University in Durham, North Carolina, appears to get hard-wired during the teen years.

Evidence from some studies also suggests that alcohol is more likely to damage memory and learning ability in the hippocampus of the evolving adolescent brain. At the same time, adolescent rats—whose brains are relatively similar to those of adolescent humans—suffer less from some of alcohol's other effects, including sedation. That sounds like a good thing, but if it is indeed true for adolescents (and for obvious ethical reasons researchers don't put adolescents through alcohol-related trials), it means they can drink more, and for longer periods—and therefore run a greater risk of long-term damage. Repeated alcohol use during these years may also lead to lasting memory and learning impairment—not to mention the fact that young binge drinkers are more likely to set themselves up with a lifetime alcohol-abuse problem.

This is one area where brain-research findings have affected how Giedd, the father of four, behaves as a parent. "In terms of substance abuse and alcohol, I'm a lot less hip now," he says. "I wouldn't have the mentality of, 'Oh it's better to have them do it at home.' [Adolescence is] a very vulnerable time in brain development to be exposed to these other substances." Giedd is surprised by how many parents say that their kids are going to drink and take drugs anyway, so they might as well do it at home, in a safe environment. "Biologically, it's a time when the cement is setting. If people cannot do these things until the age of nineteen, the odds of them not having trouble as adults go up enormously."

But experimenting, taking risks, and searching for good times are, it would seem, all part of the adolescent picture. As difficult as it is for parents to grasp, adolescents don't always make poor choices just to get their goats, or because they're suddenly gripped by temporary insanity. This sort of behaviour appears to be a predictable part of the identity-formation process, which begins in the early years but dramatically accelerates during adolescence. That's when children begin playing different roles, trying on different hats, figuring out if they're gay, straight, or bisexual, whether they're a geek, a jock, or cool. At the same time, their frontal lobes aren't fully developed, which means that the appetite for experimentation doesn't necessarily go along with the capacity to make sound judgments or to see into the not-so-distant future. In other words, by their very nature, teenagers are not especially focused on, or equipped to assess, the consequences of their actions.

A 2004 MRI study suggested that adolescent brains are less active than those of adults in regions that motivate reward-based behaviour. James Bjork, a neuroscientist at the National Institute on Alcohol Abuse and Alcoholism, and his colleagues conducted a brain scan on twelve adolescents between the ages of twelve and seventeen and a dozen adults aged twenty-two to twenty-eight. During the scan, the participants responded to targets on a screen by pressing a button; the object was to win (or avoid losing) varying amounts of money. The researchers found that areas of the brain associated with seeking gain lit up in both age groups. But in the adolescents, there was less activity. Adults, says Bjork, may have developed circuitry that enables them to motivate themselves to earn relatively modest rewards—the satisfaction felt after volunteering at church, say, or walking through a ravine. Adolescents, on the other hand, "may need activities that either have a very high thrill payoff or reduced effort requirement or a combination of the two." Examples, he adds, would be "sitting on the couch playing violent video games or sitting on the couch and pounding alcohol."

Even if, in quiet conversation, teenagers understand the risks of certain actions—drinking and driving, sex without protection, jumping off cliffs—when the moment of truth actually arrives, reason can be shot to hell. In the heat of the moment, the limbic area of the brain lights up like a pinball machine while the prefrontal cortex, the good angel that tamps down intense feeling and helps us navigate through emotional situations, is essentially asleep. In addition, experts have found that teenagers have a higher level of dopamine, a neurotransmitter connected to pleasure, movement, and sexual desire, which may increase the need for extra stimulation through risk-taking.

Some teenagers slide through adolescence unscathed. But there's no doubt that adolescents in the throes of hormone surges and brain development are extremely vulnerable—to making poor choices, to mental-health problems, to death and injury. A quick look at the statistics paints a troubling picture. According to Statistics Canada, adolescents between fourteen and nineteen are more likely to commit property crimes and violent offences than any other age group; 25 percent of teenagers reported binge drinking at least once a month in 2000-2001, a rate second only to the twenty- to thirty-four-year-olds. During that same period, the pregnancy rate for girls between fifteen and nineteen was thirty-six out of 1,000. Most discouraging is the suicide rate for teenagers: currently about eighteen for every 100,000, with the highest rate occurring among teenaged boys (although girls are hospitalized for attempted suicide at a far greater rate than boys).

In fact, the three leading causes of death for teenagers in North America are accidents, suicide, and homicide. Unsurprisingly, the majority of accidents involve motor vehicles; in 2004, in the United States, about 20 percent of accidents that resulted in fatalities were due to a driver who had a high blood-alcohol level. According to the Insurance Institute for Highway Safety, injuries suffered by teenagers in car crashes have become a pressing public-health problem. Sixty-two percent of teenage passenger deaths in 2004 occurred when another teenager was driving. And teenage drivers are more likely to be at fault in crashes.

All of which is not going to make parents sleep any better—if indeed they can get to sleep in the first place.

Teenage speeding, irresponsibility, and status-seeking are not the only explanations for the statistics (though teenagers have been shown to take greater risks behind the wheel when their friends are with them). In fact, they also appear to be at a disadvantage because they have not refined the ability to multitask—driving while drinking a beverage, listening to music, talking on a cellphone, or even chatting with a passenger. One sensible response to this, according to many scientists and policy-makers, is graduated licensing, which is already in place everywhere in Canada except Nunavut. In 1996, many American states started to introduce some aspects of graduated licensing, and according to a 2003 report in the *Journal of Safety Research*, they have seen a decrease in crash rates.

So if adolescents are a work-in-progress in terms of judgment, should they be held accountable for their crimes in the same way adults are? Recent adolescent brain development research was used in arguments against the juvenile death penalty in the United States. If adolescents aren't yet fully capable of controlling their emotional responses or understanding the consequences of their actions, groups like the International Justice Project said, then they should not be punishable by death. In March 2005, when the US Supreme Court finally abolished the juvenile death penalty, there were seventy-three people on death rows across the United

States for crimes they had committed before the age of eighteen. Many brain researchers believe that science should be part of the debate. But, Giedd adds, "it becomes a very slippery slope: the same data that might support abolishing the juvenile death penalty could be used to take away teenagers' ability to make their own reproductive-rights decisions."

Despite these new findings, has brain science told us anything we don't already know? Bjork's answer: "As Jay Giedd says, a lot of what we're finding out in brain research is the neuroanatomical, neurometabolic correlate of what grandma always told you." Indeed, brain mapping has provided proof of a neurological and biological basis for what sometimes ails the still-forming adult (and the adults who love and live with them).

Of great urgency for Giedd and others now is why certain disorders—anxiety and eating disorders, substance abuse, schizophrenia—develop during adolescence, but not autism, ADHD, Alzheimer's, and others. "Many of the things that plague adults really do hap- pen during the teen years," says Giedd, "so identifying them early, treating them early, when the brain is more plastic, would seem to make more sense in terms of really having a lifelong impact." Parents are wise not to assume that misery and anxiety are just part of the teenage rite of passage; it may be that serious unhappiness in adolescence is an early-warning sign of adult disorders.

Another task for scientists is to determine which things in a teenager's environment and experience will, for better or worse, influence brain development. "So many things have already been put forward—music, education in general, learning a second language, bacteria, viruses, video games, diet, sleep, exercise," says Giedd, "and all of them are probably true to some extent."

But what the general culture has to offer to teenagers is only one part of the equation. The brain has always been built for learning by example and experience—which experiences lead to pain, which lead to good outcomes. And for Giedd, that facility is what will give adolescents the best chance to grow up well—the ability to learn from the people around them. "It's the little things, the day-to-day things that we say in the car or when we're solving problems, how we handle relationships, emotions, our work ethic," he says. "They will believe much more what we do than what we tell them."

In fact, if there is anything parents can take away from all the scientific research into adolescent brain development, it's that their influence, patience, understanding, and guidance are very necessary—even when the teenager or young adult shrinks away from affection, grunts, slams doors, blasts music, rolls eyes, breaks house rules, and seems incapable of following simple instructions. Developing brains often can't handle organizational problems; they have more trouble making social, political, and moral judgments; they have to be reminded of potential consequences and carefully directed toward risks that aren't quite so, well, risky. Developing adults need appropriate amounts of independence, freedom, and responsibility.

"I would say with a clear conscience that the teen brain is different than the adult brain," says Giedd. "Just as I would feel comfortable saying men are taller than women." We ignore those differences at our peril, he adds. Teenagers may drive the family car, move away from home, go to college, and spend their early twenties wrestling with life decisions, all of which are a normal part of growing up. But as Giedd says, just because adolescents have left childhood behind, "parents shouldn't say, 'My work is done.'"

When Does Your Brain Stop Making New Neurons?[*]

A. Infant; B. 42 Years Old; C. 53 Years Old

Answer: None of the Above

By Sharon Begley
Newsweek, June 24 2007

The scientists are not so naive as to think they have discovered a magic wand that can turn animosity into compassion and hatred into benevolence, but the tarantula definitely raised their hopes. Over the years psychologists Phillip Shaver and Mario Mikulincer had uncovered more and more evidence that people's sense of emotional security shapes whether they become altruistic or selfish, tolerant or xenophobic, open or defensive. Once upon a time, that would have been that, for whatever their roots such traits were thought to be, by adulthood, as hard-wired as a computer's motherboard.

But with the new millennium scientists were finding that brain wiring can change, even in adults. That got Shaver, a professor at the University of California, Davis, and Mikulincer, at Israel's Bar-Ilan University, thinking: could they activate unused or dormant circuits to trigger the sense of emotional security that underlies compassion and benevolence? To find out, they gave volunteers overt or subliminal cues to activate brain circuitry encoding thoughts of someone who offered unconditional love and protection—a parent, a lover, God. The goal was to induce the feeling of security that makes it more likely someone will display, say, altruism and not selfishness. It worked. People became more willing to give blood and do volunteer work, and less hostile to ethnic groups different from their own. Offered a chance to inflict pain on an Israeli Arab with whom they were paired in an experiment (serving him painfully spicy hot sauce), Israeli Jews did not dole it out as they did without the security-circuit activation. They held back. And when they saw a young woman distraught over having to pick up a tarantula as part of an experiment, they volunteered to take her place.

OK, so they didn't all sign up to work in Darfur. But as recently as a decade ago, proposing that an adult brain could be rewired for compassion—or anything else, for that matter—without experiencing a life-altering epiphany would have been career suicide for a neuroscientist. Not anymore. Experts are overthrowing the old dogma that, by the ripe old age of 3, the human brain is relatively fixed in form and function. Yes, new memories could form, new skills could be mastered and wisdom could (in some) be gained. But the basic cartography of the adult brain was thought to be as immutable as the color of your eyes. This "neurological nihilism," as psychiatrist Norman Doidge calls it in his recent book, "The Brain That Changes Itself," "spread through our culture, even stunting our overall view of human nature. Since the brain could not change, human nature, which emerges from it, seemed necessarily fixed and unalterable as well."

But the dogma is wrong, the nihilism groundless. In the last few years neuroscientists have dismantled it pillar by pillar, with profound implications for our view of what it means to be human. "These discoveries change everything about how we should think of ourselves, who we are and how we get to be that way," says neuroscientist Michael Merzenich of the University of California, San Francisco. "We now know that the qualities that define us at one moment in time come from experiences that shape the physical and functional brain, and that continue to shape it as long as we live."

The brain remains a work in progress even on so basic a parameter as its allotment of neurons. For decades, scientists assumed that adding new neurons to this intricate machine could only spell trouble, like throwing a few extra wires into the guts of your iPod. But in 1998 Peter Eriksson of Sweden's Sahlgrenska University Hospital and colleagues discovered that brains well into their 60s and 70s undergo "neurogenesis." The new neurons appear in the hippocampus, the structure deep in the brain that takes thoughts and perceptions and turns them into durable memories. And studies in lab animals show that the new neurons slip into existing brain circuits as smoothly as the Beckhams onto the Hollywood A list.

Brain structure is also malleable, recording the footprints of our lives and thoughts. The amount of neural real estate devoted to a task, such as playing the violin, expands with use. And when an area of the brain is injured, as in a stroke, a different region—often on the mirror-image side—can take over its function. That overthrew the long-held view called "localizationism," which dates back to 1861, when French surgeon Paul Broca linked the ability to speak to a spot in the left frontal lobe. But contrary to the belief that particular regions are unalterably wired for specific functions, even one as basic as the visual cortex can undergo a career switch. In people who lose their sight at a young age, the visual cortex processes touch or sound or language. Receiving no signals from the eyes, the visual cortex snaps out of its "waiting for Godot" funk and reactivates dormant wires, enabling it to perform a different job.

If something as fundamental as the visual cortex can shrug off its genetic destiny, it should come as little surprise that other brain circuits can, too. A circuit whose hyperactivity causes obsessive-compulsive disorder can be quieted with psychotherapy. Patterns of activity that underlie depression can be shifted when patients learn to think about their sad thoughts differently. Circuits too sluggish to perceive some speech sounds (staccato ones such as the sound of "d" or "p") can be trained to do so, helping kids overcome dyslexia. For these and other brain changes, change is always easier in youth, but the window of opportunity never slams shut.

From these successes, neuroscientists have extracted a powerful lesson. If they can identify what has gone wrong in the brain to cause, say, dyslexia, they might be able to straighten out aberrant wiring, quiet an overactive circuit or juice up a sluggish one. It won't happen overnight. But UCSF's Merzenich believes we have glimpsed only the surface of the ability of the brain to change. "The qualities that define a person have a neurological residence and are malleable," he says. "We know that in a psychopath, there is no activation of brain areas associated with empathy when he sees someone suffering. Can we change that? I don't know exactly how, but I believe we can. I believe that just as you can take a 17-year-old and put him through basic training, inuring him to violence, we can take a person who is insensitive and make him sensitive to others' pain. These things that define us, I'm convinced, can be altered." Only more research—and it's coming—will reveal how easily, and how much.

But what of the genes that shape our disposition and temperament? Here, too, malleability rules. As is often the case, this effect is easiest to detect in lab animals. Rats develop starkly different personalities depending on how they are reared. Specifically, if Mom is attentive and regularly licks and grooms them, they become well-adjusted little rodents, mellow and curious and non-neurotic mouse or rat. If Mom is neglectful, her pups grow up to be timid, jumpy and stressed out. Once, this was attributed to the powerful social effects of maternal care. But it turns out that Mom's ministrations can reach into the pups' very DNA. Maternal neglect silences genes for receptors in the pups' brains, with the result that they have few receptors and hence a hair-trigger stress response. Maternal care keeps these genes on, so the pups' brains have lots of receptors and a muted stress response. Inattentive moms also silence the genes for estrogen receptors in their daughters' brains; the females grow up to be less attentive mothers themselves. "It's almost Lamarckian," says Francis Champagne of Columbia University, referring to the discredited idea that offspring can pass along traits they acquire during life. "But experiences during a lifetime are passed on to the next generation."

Scientists are now beginning to see the first glimmerings of this in people, too. Very young children born with the form of a gene called 5-HTT associated with shyness usually are quiet and introverted. But by age 7, scientists led by Nathan Fox of the University of Maryland find, many are not. Only if the children have certain experiences—best guess: being raised by a stressed mother unable to provide emotional and physical protection—does the "shyness gene" live up to its billing. The molecular mechanism by which experiences reach down into the double helix and inhibit or elicit the expression of a gene is not as clear in people as it is in lab rats. At least, not yet. But it's an early sign that we are not necessarily slaves to the genes we inherit.

Few laypeople understand that neurological nihilism and genetic determinism have been so discredited. Most still embrace the idea that our fate is written in our DNA, through the intermediary of the brain wiring that DNA specifies. "It's puzzling that determinism is so attractive to so many people," says UCSF's Merzenich. "Maybe it's appealing to view yourself as a defined entity and your fate as determined. Maybe it's in our nature to accept our condition."

There is an irony to that. When people believe that their abilities and traits are fixed, interventions meant to improve academic performance or qualities such as resilience and openness to new experiences have little effect. "But if you tell people that their brain can change, it galvanizes them," says psychologist Carol Dweck of Stanford University, whose 2006 book "Mindsets" explores the power of belief to alter personality and other traits. "You see a rapid improvement in things like motivation and grades, or in resilience in the face of setbacks." None of that happens, or at least not as readily, in people who believe they are stuck with the brain they have.

This is not to say that everything will yield to the new brain science. There may turn out to be aspects of ourselves that resist every effort at change, for which we may be glad. But for too many decades, science sold the brain short. It is way too early in the battle against neuro-nihilism to declare anything beyond the reach of the brain's potential to transform itself.

4

How the Brain Processes Language

Editor's Introduction

Of all the things that separate man from beast, perhaps none is more fundamental than language. While many animals communicate using sound, humans alone have the capacity to form words, organize them into sentences, and give voice to ideas that carry both literal and figurative meaning. Scientists have long sought to determine which parts of the brain are responsible for language, and in the days before neuroimaging, they gathered much of their information by studying patients with different types of head injuries. Based on these early tests, researchers developed what the psychologist Gary Marcus calls a "Swiss Army Knife view of the brain," one in which specific regions perform specific tasks. For years, researchers believed that the brain processes language in a sequential fashion: Upon hearing a sentence, it first uses the left hemisphere to try to determine literal meaning. In the case of basic statements, such as, "It was hot outside," processing might stop there. For more complex, figurative sentences, however, the theory was that the right hemisphere would then take over. "It was a volcano outside," for example, might necessitate right-brain thinking, since its meaning isn't literal. Thanks to fMRI and other neuroimaging techniques popularized during the "Decade of the Brain," scientists are developing a clearer picture of how the brain tackles language. While there appears to be some validity to the classical right-brain, left-brain hypotheses, recent findings suggest the distinctions aren't quite so simple.

The articles in this chapter provide an overview of what advanced neuroimaging has taught scientists about how the brain handles language. In the first selection, "Addressing Literacy Through Neuroscience," Steve Miller and Paula A. Tallal discuss how neuroimaging has led to breakthroughs in "brain-based" teaching methods, which educators are using to help children with dyslexia and other language deficiencies. Many children who struggle with language have difficulty processing sounds, and through the use of computer programs such as Fast ForWord, which focuses on acoustic cues, teachers are able to take advantage of the brain's plasticity and "normalize" the differences that cause some students to lag behind. The authors cite one study in which dyslexic children, after eight weeks of training, were able to score in the normal range on standardized reading tests.

In his piece "Mapping Metaphor: This is Your Brain on Figurative Language," Kenneth W. Krause explains why scientists are rethinking "modularity," or the

"Swiss Army Knife" model of language processing. While neuroimaging supports the idea that the right hemisphere plays an important role in processing figurative language, studies have shown that the left hemisphere is also active. Further, some researchers have abandoned the sequential model in favor of "coarse coding," which maintains that sentences with a "close semantic relationship" will automatically activate the left brain, while more complex sentences will activate the right, regardless of whether the meaning is literal or figurative. In the next selection, "The Correlation Between Brain Development, Language Acquisition, and Cognition," Leslie Haley Wasserman examines how children develop language, highlighting the "pre-language" stage, which lasts for 10 to 12 months after birth, and the linguistic stage, which continues for another two years. Wasserman explains that children begin life with the necessary wiring to learn any language, and that if they are not exposed to language as infants, they may have difficulty learning to speak later in life.

In the final article, "Brains Show Two Sides of Language Function," Bruce Bower reports that some people have advanced language capabilities in both hemispheres of their brains, instead of just the left, as is most common, or right. These individuals are able to retain verbal skills even after one side of the brain has been damaged, according to findings. The same study suggests that people dependent on either the right or left hemisphere are better able to match words with pictures when the non-dominant halves of their brains are temporarily suppressed. "The nature of language representation in the brain still remains unclear," Glyn W. Humphreys, a psychologist at the University of Birmingham, tells Bower.

Addressing Literacy Through Neuroscience[*]

By Steve Miller and Paula A. Tallal
School Administrator, December 2006

What is he thinking?

It's a typical Monday afternoon and you are talking with a teenage boy about a short story he read in school earlier that day. He seems to be having unusual difficulty expressing his thoughts about the story. Did we read the same story? Why is this so difficult for him? What's going on inside his head?

Most educators have asked themselves the latter question more than once. And they are right to focus on the head—or more precisely the brain—as it holds the secrets to understanding human behavior.

The brain is the source of all of our thoughts, feelings and emotions. Much of what defines us as humans—our creativity, our use of language, our critical thinking skills—is housed somewhat mysteriously in the brain. Now the mysteries of the human brain are rapidly being elucidated by neuroscience research. In the past few decades, no area of importance to educators has been so significantly transformed than our understanding of the way the brain learns.

NEW INSIGHTS

For more than 150 years, neuroscience has held that most of the brain's functionality develops during critical periods in early childhood and that once past these critical periods, the window of opportunity for brain modification slams shut. However, today, after decades of research, we've abandoned this view of the brain as analogous to a hard-wired computer in favor of the concept that the brain is a continuously modifiable "plastic" organ throughout life.

This realization provides new opportunities for us to explore and develop a more complete understanding of human brain development and cognition and, most importantly to educators, to create neuroscience-informed instructional strategies that enhance the brain's capacity to modify itself through learning, known as neuroplasticity.

The brain learns at the physiological level of the neurons by looking for consistencies in what we experience, learning to attend to and map those patterns and events that repeat themselves frequently. Those events are usually made up of sensory input coming into the brain from the five senses. The brain's job, beginning at birth, is to code neurally, based on its experiences, what's going to matter and what's not going to matter and also to predict what is going to happen next. This form of learning is known as experience-dependent learning or neuroplasticity.

Throughout our lives, as we learn new skills and acquire new knowledge, our brains are continuously being physiologically remodeled, creating selective clusters of brain cells, or neuronal cell assemblies, that respond more and more automatically and synchronously in time to events that come together or follow one another closely in time. Put simply, neurons that fire together wire together. That is, the more often a specific pattern occurs (such as a pattern of acoustic changes that occurs within speech), the more likely that pattern will be neurally coded or represented for easier and more efficient access at a later time.

This fundamental discovery of the role of experience-dependent learning, and specifically the precise scientific learning principles needed to enhance cognitive capacity, has led to major advances in the neuroscience of learning. Over the past 10 years, with the aid of computer technology and the Internet, these advances have been translated out of the research laboratory into classrooms nationwide to help students struggling with spoken and written language skills. This brain-based approach to intervention for struggling students has been shown to rapidly and efficiently enhance the fundamental learning skills (memory, attention, processing and sequencing) that form the foundation upon which efficient learning depends, specifically learning language and reading.

LANGUAGE TO LITERACY

Like other complex tasks, reading is not an innate skill that develops spontaneously. Rather, it must be taught, practiced and learned. But before children can learn to read proficiently, they must first be able to understand and produce spoken language in the same language they are learning to read. It's not that you can't learn to read if you are not sufficiently proficient in the language you are trying to learn to read, but it's very difficult.

This language-to-literacy link compels us to explore the components of spoken language, the organization of these components in the brain and the links between them and reading so we can understand how the brain learns to read. Language is comprised of five basic components: phonology, morphology, semantics, syntax and pragmatics. The first four are essential components of the reading process as well.

PHONOLOGY

Phonemes are the building blocks we use to construct words. They are the smallest unit of sound in spoken language that can change the meaning of a word. For example, the word big has three phonemes—/b/, /i/ and /g/. Changing one of the phonemes changes the meaning of the word, such as changing the /b/ to /d/ and producing dig. English has 44 different phonemes. When the acoustic patterns that make up phonemes repeatedly enter the brain as the infant listens to speech, the brain represents them as an assembly of brain cells (neurons) that fire together closely in time.

These neural representations of phonemes play an important role in learning to talk and subsequently in learning to read. Basically, the brain learns these distinct sounds and packages them as a cluster of neurons that fire in a certain pattern. When learning to read, the child must become aware that it is these patterns that must be extracted from inside of words and attached to letters (graphemes). This is known as phonemic awareness.

It is important to recognize that children are born with the ability to process the phonemes of all languages. After all, the brain does not know which language it will have to learn until it is exposed to it. This is where experience-dependent learning becomes critical. As infants listen to the language(s) spoken around them in the first year of life, their brain forms connections for only the phonemes they hear consistently—those of their native language. Phonemes that do not occur in that language or are too difficult for some reason for an infant to process are not wired into the brain. This has important consequences for understanding the neurobiological basis for reading difficulties later in life.

Following the rapid acoustic changes within the ongoing speech stream, which are critical to discriminating between the phonemes in a language, is one of the fastest things the human brain has to do. For example, only 40 milliseconds at the onset of the word big distinguishes it from the word dig.

Research has shown that for a variety of genetic as well as environmental reasons, many children who struggle with spoken and written language have difficulty tracking acoustic changes that occur this quickly. This interferes with their ability to map the distinct phonemes of their native language needed in both spoken and written language development. Importantly, if infants do not map distinct representations of each phoneme in the same language that they will ultimately be taught to read (either through lack of experience with that language or because of other factors that create a roadblock in the brain), they will struggle to retrieve the sounds inside of words that must be attached to letters in order to become a proficient reader.

The brains of children who are hearing impaired or have central auditory processing problems have difficulty representing phonemes correctly because their brain does not receive accurate acoustic input. As a result, these children are at risk for developmental language learning impairments, which can include both spoken and written language.

The same is true of children who are not native speakers of English. They do not have all of the English phonemes represented in their brain because they were not "wired in" through their early language experience. Children who struggle to learn to read English can benefit from training programs that are designed explicitly to emphasize the acoustic differences between English phonemes while also strengthening the other components of spoken English (morphology, semantics, syntax) that form the foundation of written English.

One of the most important ways in which research on neuroplasticity has been translated into the classroom is through the development of novel computer-based intervention programs, such as the Fast ForWord series of language and reading programs that have been shown to improve the rate of acoustic processing and sharpen phoneme perception and phonemic awareness, all critical to reading.

These programs are unique in that they explicitly were designed to employ the scientific learning principles underlying neuroplasticity (which include frequency/intensity of input, individual adaptability, sustained attention and timely reward) to enhance the fundamental memory, attention, processing and sequencing skills on which effective classroom learning depends. Controlled scientific studies have been conducted that demonstrate that these neuroplasticity-based training programs lead to rapid and sustained improvements in English language skills as well as reading skills in struggling learners. (See www.scientificlearning.com/results.)

MORPHOLOGY AND SEMANTICS

Morphemes are the smallest unit of meaning in language. Free morphemes are root words such as cat. Bound morphemes are suffixes, like an "-s" added to a root morpheme to create a plural, and prefixes, like an "un-" added to create an antonym of the root word. Free morphemes can stand alone; bound morphemes must bind to a root word to make sense.

By middle school, morphological awareness becomes more important than phonemic awareness for reading comprehension and for fluent, proficient reading. For children to become better readers in middle school, they may benefit most from programs that have been developed explicitly to enhance their knowledge of multimorphemic words (words such as unnaturally or distastefully).

Semantics rescues us from gibberish by providing a set of common vocabulary words and definitions that all native speakers of a particular language use to refer to objects or concepts. When we hear one of these words, we look it up in our brain's "mental dictionary" to retrieve its meaning.

Semantics also includes our brain's organization of word categorization systems. Word meanings include knowledge not only of an individual word, but also how that word relates to other words. For example, cat can be categorized in the most general sense as an animal. But it also can be categorized specifically as an animal with four legs and even more specifically as a household pet.

Words are stored in the brain together with other words that are related to them phonologically, morphologically and semantically. Interestingly, damage to specific areas of the brain can result in the loss of one category of words (such as all fruit), but spare others (such as all animals). Another interesting finding from neuroscience research is that when two languages are learned simultaneously very early in development, they overlap in the language areas of the brain. However, when one language is learned before another is begun, then they occupy discrete areas next door to each other in the language areas of the brain.

These insights from neuroscience research help inform the development of computer-based training exercises that aim at improving the semantic skills of children learning English as a second language as well as struggling readers.

SYNTAX

Syntax refers to the part of speech of a word (for instance, noun or adverb), the grammar of language as expressed by word order and grammatical morphemes, and the set of rules for combining words into sentences. Grammar helps us clarify meaning by providing a set of rules for creating explicit relationships among words. We use these rules to indicate exactly who is doing what and to whom. It also allows us to comprehend ambiguous words based on context. For example, we know, based on syntax, whether the word bark is referring to a part of a tree or a dog's vocalization. Syntax provides structure for almost everything that we say, write, read or hear.

To read English proficiently, children must understand the grammatical rules of spoken English. This can be particularly challenging when working with children whose native language is not English. Sometimes a student's incomplete knowledge of English is due to a lack of sufficient environmental input (children from low socioeconomic families are exposed to far fewer words than those from professional families) or due to a feature of the student's dialect that causes syntactical errors. But sometimes these errors result from the brain's inability to process sounds quickly enough.

Many grammatical endings in English, such as "-s" for third-person singular verbs or "-ed" for the past tense, are exceptionally brief in sound duration, especially when they occur within a sentence. Many children who are struggling to learn to read have difficulty picking up on the rapidly successive sounds within words, including these brief grammatical morphemes. When a student's neural processing of sounds is slow, the result may be phonological, morphological and/or syntactical errors in spoken and written language.

The Fast ForWord series of language and reading training programs explicitly focuses on increasing students' brain processing speed. These programs also use a patented speech algorithm that finds the brief segments within the ongoing speech stream (be they within phonemes, morphemes or syntax) and enhances these acoustic cues by making them longer and louder. As students progress in phonological, morphological, semantic and syntactic skills, the amount of acoustic modification decreases until they can perform adequately in each of these areas with normal speech.

THE READING BRAIN

When you are familiar with the basic elements of spoken and written language, it's easier to understand what is happening in your students' brains as they gather information about phonology, morphology, semantics and syntax when listening to or reading sentences or paragraphs.

As the electrical impulses representing each phoneme, morpheme or word course through the temporal lobe and into the frontal lobe of the brain, the brain picks up meaning and associations about the words (semantics) from areas of the temporal cortex. When a student processes grammatical morphemes, her superior temporal gyrus activates. When a student determines whether a word is a noun or a verb, his frontal and temporal regions activate.

Before reading can map onto the language areas of the left hemisphere, the occipital lobe, located in the back of the brain, needs to be activated. The occipital lobe is primarily responsible for vision-related functions and, therefore, must process the visual features of the letters. The brain must learn to recognize the visual form of each letter and store a mental representation based on experience-dependent learning similar to that used for learning representations of each phoneme.

Reading comes together in the temporo-parietal cortex in the left hemisphere, located above and behind the left ear. This is where information about letter shape, word recognition, meaning and sound is integrated in the brain. Damage to this area of the brain can affect reading ability in children or adults.

REWIRING TO READ

Students' struggles to read can be caused by any number of factors, including genetic predisposition (language and reading disabilities often run in families), middle ear infections that may lead to blocking sound intermittently getting into the brain when phonemes are being mapped, lack of adequate and consistent exposure to words in the home during early development and/or the language used in the home is not the same as that used at school or for reading.

However, research on neuroplasticity has opened the door for developing novel, neuroscience-informed methods for enhancing basic cognitive and linguistic skills that are critical to reading success. This raises the question of whether the brains of struggling readers actually can be "rewired" to function more like those of normal readers?

To find the answer, Elise Temple, an assistant professor of human ecology, and her colleagues at Cornell University, conducted a study to determine whether the lack of activity in the temporo-parietal cortex of dyslexic readers could be normalized by behavioral training.

The study used the training program Fast ForWord Language. It focuses on auditory processing and spoken language through an intensive and adaptive series of computer exercises. One unique feature of the program is a focus on training the basic acoustic and cognitive skills (memory, attention, processing and sequencing) that are critical to listening comprehension, phonological and morphological awareness and other aspects of both spoken language and reading. The series of computer exercises, designed to mimic experience-dependent learning in the brain, specifically develops the components of spoken language that are most essential for reading.

In the study, children 8 to 12 years old with dyslexia underwent fMRI scans before and after eight weeks of participation. A control group of normal readers underwent two fMRI scans about eight weeks apart to control for practice effects. Both groups performed a phonological processing (letter rhyming) task while undergoing fMRI. Results were compared to a letter-matching task in which the child simply indicated whether two letters were the same or different. By comparing the brain function during the rhyming task with the brain function during the matching task, the researchers could observe the part of the brain that was activated for phonological analysis rather than orthographic (visual) processing of letters.

The results of the study were published in the Feb. 25, 2003, issue of the Proceedings of the National Academy of Sciences. During the first fMRI scan before intervention, the dyslexics showed the expected lack of activation in their temporo-parietal cortex compared to normal readers. Also as expected, the dyslexics' performance on standardized reading tests was outside the normal range for their age. After eight weeks of participation, the dyslexic children's performance on standardized reading tests improved significantly, moving into the normal range in all areas (word identification, word attack and passage comprehension). Increased activation in the left temporo-parietal cortex was also evident.

This is the first study to use fMRI to document scientifically that the brain differences seen in dyslexics can be "normalized" by neuroplasticity-based training. Perhaps of greater relevance to educators, parents and the children themselves are the accompanying significant increases in reading scores on standardized tests that were also documented as the result of this intervention.

LIFELONG PROCESSING

Brain research shows us that literacy problems are not simply a matter of a child not trying hard enough or of poor instruction. Nor are they likely to be solved by providing more traditional reading instruction. Rather, this line of research demonstrates that the biological as well as the experience-dependent aspects of learning to talk and to read require a neuroscience-informed approach to intervention for those students who are failing to make adequate yearly progress using more traditional methods alone.

The exciting news is that the brain remains open to neuroplastic modification throughout life when education and neuroscience work together to attack literacy problems. A third leg of this partnership is the use of advanced technology that allows laboratory research not only to be translated effectively for individualized instruction, but also scaled up so that it can be implemented broadly in a variety of K-12 school settings in a reliable, efficient and cost-effective manner. Because these neuroscience-based discoveries were translated initially for classroom use, they have been made available to more than 750,000 students in 4,000 schools nationwide.

Mapping Metaphor[*]

This Is Your Brain on Figurative Language

By Kenneth W. Krause
The Humanist, July/August 2008

PORTRAIT

> "Children may not understand political alliances or intellectual argumentation, but they surely understand rubber bands and fistfights."—Steven Pinker, from *The Stuff of Thought: Language as a Window into Human Nature* (Viking, 2007).

Sometimes a cigar is just a cigar. Then again, mischief is the hot smoke that curls off the end of a lit intellect. And sometimes a diamond in the rough is indeed just an ancient deposit of highly compressed carbon. But no facet of humanity's evolved "genius," as Aristotle put it more than 2,300 years ago, sparkles so brilliantly as our unique capacities for extra-literal description and comprehension.

Until recently, most professional sources have attributed our proficiency with language to a pair of knuckle-sized regions on the brain's left side called Broca's area and Wernicke's area. The former module was responsible for grammar, the latter for word meanings. But new technologies, featuring functional Magnetic Resonance Imaging (fMRI), have allowed scientists to probe non-invasively into the brains of healthy volunteers and to discover, first, that other parts of the brain share in these responsibilities and, second, that Broca's and Wernicke's regions contribute to other important tasks as well.

For some, such frustrating complexity spells the end of the "modularity" hypothesis that instructs the functional specialization of certain identifiable neural systems. For others, including psychologist Gary Marcus, author of *The Birth of the Mind* (Basic Books, 2004), new evidence suggests "not that we should abandon modules (the Swiss Army Knife view of the brain) but that we should rethink them—in light of evolution."

While he acknowledges that the brain's left hemisphere appears to be devoted to both language and problem solving, R. Grant Steen, psychologist, neurophysiologist, and author of *The Evolving Brain* (Prometheus, 2007), agrees with Marcus, defining language (as opposed to mere communication) in very practical, adaptive terms:

> [L]anguage is a system of communication that enables one to understand, predict, and influence the action of others. Inherent in this definition is a concept of theory of mind: if communication is instinctual rather than having a purpose, then it should probably not be considered a language. If communication has a purpose, this assumes an awareness of other independent actors, whose actions can potentially be influenced. . . . [F]or communication to serve the needs of the listener as well as the needs of the speaker, the listener must be able to understand what the speaker is "really" saying. It is not enough to understand the literal meaning of speech.

Broadly stated, then, experts seek out the neural substrates and processes of figurative language comprehension in order to distinguish the biological bases of what makes humans most exceptional among animals. In the end, they hope as well to develop more effective means of restoring these extraordinary abilities to those who have lost them and, perhaps, to enhance such talents for the benefit of our collective future. Although the search has just begun, we have already learned a great deal.

Two intimately associated paradigms have suffered intense scrutiny in recent years. The standard model of figurative language processing—sometimes referred to as the "indirect" or "sequential" view—maintains that the brain initially analyzes passages for literal meaning and, if the literal interpretation makes no sense, then reprocesses the words for access to an appropriate figurative meaning. According to the related dichotomous model of "laterality," the brain's left hemisphere (LH) is responsible for processing literal language while its right hemisphere (RH) is enlisted only to decode figurative expressions.

Such paradigms were based on classic lesion studies beginning with those conducted in 1977 by Ellen Winner and Howard Gardner who showed that patients with RH damage had much more difficulty processing metaphors than subjects with LH damage. However, in an editorial from the February 2007 issue of *Brain and Language*, linguistics expert Rachel Giora argued that Winner and Gardner's results had been widely misinterpreted. Although only the LH patients in the lesion studies were able to competently match metaphorical figures with their corresponding pictures, Giora explained, it was not true that the RH patients were unable to make such connections when asked to do so verbally. Indeed, a number of studies published at the turn of the century challenged the notion that RH damage selectively impairs people's command over verbal figurative language.

During the last few years, researchers have begun to dissect the old paradigms more systematically. In August of 2004 Alexander Rapp's team of German scientists published a report in *Cognitive Brain Research* titled "Neural Correlates of Metaphor Processing." They used event-related fMRI technology to detect brain activity in sixteen healthy subjects as they read short, simple sentences with either a literal or a metaphorical meaning.

Consistent with the laterality model, Rapp had predicted that metaphorical versus literal sentences would induce more vigorous brain activity in participants' right lateral temporal cortices. Instead, the strongest signal disparities occurred in the subjects' LH, the left inferior frontal and temporal gyri, or cortical folds, in particular. In possible contradiction to the indirect or sequential view of metaphor processing, Rapp's study noted as well that neither response times nor accuracy diverged between the two conditions. In summary, the team advised their colleagues to reassess the RH theory of figurative language comprehension and posited that, although the RH appeared to play some important role, factors other than figurativity per se might be involved.

Two years later, cognitive scientists Zohar Eviatar and Marcel Adam Just published a similar study, "Brain Correlates of Discourse Processing: An fMRI Investigation of Irony and Conventional Metaphor Comprehension" in *Neuropsychologia*. There, sixteen subjects digested ironic sentences (e.g., When Ann came home covered in mud, her mother said, "Thanks for staying so clean.") in addition to literal and simple metaphorical expressions.

As one might guess, the results were considerably more complicated. First, all three types of statements stimulated the classical language areas of the LH: moving roughly from front to back, the left inferior frontal gyrus, the left inferior temporal gyrus, and the left inferior extrastriate region. Second, metaphorical sentences activated all these same areas to a significantly higher degree than did either literal or ironic statements. Third, the right superior and middle temporal gyri were significantly more sensitive to ironic statements than to any others and, finally, the right inferior temporal gyrus was differentially sensitive to metaphorical meanings.

From these varied results, Eviatar and Just concluded that because all kinds of stimuli had activated the same classical language regions of the LH, the exclusive RH theory of figurative language as such was untenable. In addition, both metaphor and irony had triggered further brain activation—metaphor most conspicuously in the LH and less forcefully in one part of the RH, and irony quite vigorously in a rather disparate region of the RH. For whatever reasons, then, the metaphors were processed in a slightly dissimilar way than the literals. Perhaps most significantly, the metaphorical and ironical expressions were processed differently in relation to one another.

The authors proposed a number of possible causes for this last distinction, but seemed inclined to attribute it to the sentences' character rather than to their category. Recall that Eviatar and Just had chosen conventional (sometimes called "salient") metaphors. Long hackneyed, such expressions have been "lexicalized" to the point where people really don't have to think about them in order to understand them. In this experiment, for example, a fast worker was compared to a "hurricane" and a conscientious sister was likened to an "angel from heaven." Simple, idiomatic metaphors like these, the authors speculated, might be processed most efficiently in the LH as a unit, not unlike long words and literal phrases.

Irony, on the other hand, is always more interpretive and complex because it implicates an association between the speaker's thoughts and the thoughts of someone else. Citing developmental studies relating to theory of mind mechanisms, the authors alluded to the fact that, while healthy children and adults who are able to correctly attribute first order beliefs (modeling what another person knows) are also able to comprehend metaphor but not necessarily irony, subjects who can make second order attributions (modeling what another person knows about what a third person knows) are usually capable of understanding irony as well. As such, Eviatar and Just prodded, the possibility that complexity rather than figurativity per se might be responsible for RH involvement invoked "an extremely interesting set of issues for future research."

Psychologist Gwen L. Schmidt apparently concurred before she and her team of American researchers announced the results of their study, "Right Hemisphere Metaphor Processing? Characterizing the Lateralization of Semantic Processes," in the February 2007 edition of *Brain and Language*. Instead of fMRI, the authors used a divided visual field technique where the reaction times of eighty-one subjects were measured after reading the final, experimentally relevant portions of sentences either in their left visual fields (to test stimulation of the RH) or in their right visual fields (to check activation of the LH).

Three different phases were designed to investigate how the brain processes various types of figurative and literal sentences. Phases one and two compared reaction times between moderately unfamiliar (or "non-salient") metaphors (e.g., This city is a chimney) and both familiar and unfamiliar literals (e.g., The children's shoes were covered in dirt and Janice used fans axes). Phase three compared times between familiar and highly unfamiliar metaphors (e.g., Alcohol is a crutch and A bagpipe is a newborn baby). During the first two trials, the team recorded a RH processing time advantage for moderately unfamiliar metaphor sentence endings and a LH advantage for literal-familiar sentence endings. Literal-unfamiliar sentences, like novel metaphors, produced an advantage for the RH. During the final trial, the authors found a LH advantage for familiar metaphors and a RH advantage for their highly unfamiliar complements.

In other words, Schmidt and company got exactly what they had expected consistent with the "coarse coding model" of semantic processing. Displacing the old indirect/sequential processing and dichotomous laterality paradigms, the coarse coding model predicts that any sentence depending on a close semantic relationship (e.g., The camel is a desert animal) will activate the LH, and that any sentence relying on a distant semantic relationship (e.g., either The camel is a desert taxi, or The camel is a good friend) will activate the RH, regardless of whether the expression is intended metaphorically or literally. More hackneyed stimuli can be efficiently processed in a fine semantic field in the LH. Novel ones with multiple possible meanings, however, must be dealt with more methodically in a much coarser field in the RH.

All of which makes good, practical sense from an evolutionary point of view. Adaptations are cumulative, of course, and nature builds ever so slowly and imperfectly, if at all, upon existing structures. While theoretically possible, we should never assume a priori that an isolated region of the brain would take sole responsibility for any behavior or the accomplishment of any task. Recent investigations make it clear that the human brain has evolved into a highly integrated (not to mention surprisingly plastic) organ.

But why should cognitive scientists of all people agonize over literary minutia normally regarded only in university humanities departments? Generally, because the days are long past when science could be neatly segregated from "other subjects." More specifically, because significant clinical interests are at stake as well. Several patient populations reliably suffer from diminished or otherwise altered comprehension of irony, humor, metonymy, and non-salient metaphors in particular. Certain diseases, therefore, might well find their causes in brain anomalies also responsible for linguistic deficiencies. Regardless, such deficiencies surely exacerbate the existing social impairments experienced among patients overwhelmed by serious psycho- and neuropathologies.

One of the more unfortunate features of schizophrenic thought disturbance, for example, is the still mysterious problem of "concretism," the inability to grasp non-literal language. In the January 2007 issue of *NeuroImage*, Tilo Kircher's team published "Neural Correlates of Metaphor Processing in Schizophrenia," an fMRI study involving twelve subacute in- and outpatients and twelve control subjects who inspected brief sentences with either a literal or novel metaphorical connotation. Kircher's goal, of course, was to begin the process of exposing the disease's neural bases.

As predicted, all participants' brains activated in the left inferior frontal gyrus more forcefully for metaphors than for literals. With respect to metaphors only, controls clearly reacted more strongly than patients in the RH (more specifically, the right precuneus and right middle/superior temporal gyrus). LH results were more complicated. Healthy subjects activated most vigorously in the anterior portion of the left inferior frontal gyrus, a locus equivalent to what researchers call Brodmann's areas 45 and 47 (which, incidentally, is just anterior-inferior to the classical Broca's area). Remarkably, this region has been closely associated with sentence-level semantic language comprehension. By contrast, patients activated most impressively in Brodmann's area 45, three centimeters dorsal to peak stimulation among controls.

While first acknowledging prior evidence demonstrating the RH's valuable role in complex syntactic and semantic processing, Kircher's team stressed their findings that the inferior frontal and superior temporal gyri "are key regions in the neuropathology of schizophrenia," and that "[t]heir dysfunction seems to underlie the clinical symptom of concretism, reflected in the impaired understanding of non-literal, semantically complex language structures." In other words, the patients' shared failure to recruit now specifically identified areas in the LH appears to be at least pertinent if not vital to our struggle against this horribly debilitating illness.

In an even more recent edition of *Brain and Language*, a group of Italian psychologists and neurologists led by Martina Amanzio published "Metaphor Comprehension in Alzheimer's Disease: Novelty Matters," a study comparing both conventional and novel metaphor comprehension among twenty probable Alzheimer's sufferers and twenty matched controls. Based in part on some of the above-referenced experiments, Amanzio successfully predicted that patients would perform relatively well with salient metaphors but significantly less so with non-salient ones.

While maintaining a healthy skepticism, the team hypothesized that the distinction might this time involve the prefrontal cortex, the brain's executive center, because prefrontal dysfunction is a common symptom of Alzheimer's disease and because the comprehension of non-salient metaphors requires the executive ability to compare and combine vehicles and topics in order to appreciate figurative meanings. "These findings," the Italians concluded, "may have some clinical implications for the real life communication with [Alzheimer's] patients. Salience matters."

And, thus, so does metaphor. Figurative language is surely more than an intellectual extravagance. It is as much a fiber of our very being as each of the countless neurons contained in our big, beautiful brains. Most fortunately, however, comprehension of novel expression serves as a useful barometer of our personal and communal health as well. So one might permit a writer the guilty pleasure of mixing his metaphors on occasion, despite academic decorum.

The Correlation Between Brain Development, Language Acquisition, and Cognition[*]

By Leslie Haley Wasserman
Early Child Education, June 2007

ABSTRACT

There continues to be a debate whether educators should use brain research to their advantage in the classroom. This debate should not prevent educators from using their new found knowledge toward enhancing their students' learning. By understanding how the brain learns, educators are able to determine what developmental level the child is physically, mentally, socially, and cognitively. The more knowledge an educator has and applies, the better the children will learn, and our future leaders will be better educated.

Key Words: brain research; cognition; neuroscience; language acquisition; early childhood education.

INTRODUCTION

Brain research began as far back with the Egyptians in 1700 BC. With each century, brain research was improved with thinking, observation, and testing. Since the 19th and 20th centuries, the advancement of technology has caused scientists and educators to rethink and redefine brain research. The 21st century has brought about many new technological advances that help to pinpoint specific areas of the brain that have difficulty and need to be improved to aid the education of the children within our classrooms. These difficulties in cognition and learning can be diagnosed early on in the child's life. It is hoped that eventually these discoveries can lead to a decrease of disabilities in students with use of early intervention and improved technology. This new information that is being gathered about brain research can really have an impact on education and learning as more facts are uncovered with each further testing.

BRAIN CIRCUITRY

Everything a child sees, hears, thinks, and touches transfers into an electrical activity that is stored into the synapses within the brain. Each time the brain is stimulated, the experience rewires the brain. Information is carried to the brain in synapses. Each day thousands of synapses die off. Information that is not important or relevant will die off while other information that is relevant will be stored in the brain (Siegler, 2000). The brain has many synapses. Some will be preserved and others will be eliminated (Sousa, 2006). Children can lose over 20 billion synapses per day from early childhood through middle childhood and adolescence (Eliot, 2001). The brain is a complex, interconnected system that is connected to everything in the body. Due to its complexity, neuroscience has been able to delve further into the functions of the brain with the use of technology.

TECHNOLOGY AND THE BRAIN

Imaging technologies fall into two different categories: those that look at the structure of the brain and those that look at the function of the brain (Sousa, 2006). Many of these technologies were developed for medical usage such as detecting tumors. There are several different imaging technologies that can be used for diagnostic tools for educators. One of the imaging machines used for seeing how quickly something occurs in the brain is the EEG (electroencephalography) where electrodes are attached to the scalp and electronic signals from the brain are recorded (Sousa, 2006). Another of the imaging technologies that has success for determining the differing areas within the brain is the functional magnetic resonance imaging (fMRI). The fMRI is a painless, noninvasive way to pinpoint specific areas in the brain. These are just two of the many types of technologies that are being used for diagnostic testing for cognitive neuroscience as well as medical advances.

Cognitive neuroscience is using new technology to determine specific questions about the brain. With this advancement, many cognitive neuroscientists are now able to pinpoint specific areas of the brain that cause reading difficulties and language difficulties, just to name two. They are also able to pinpoint all of the different areas of the brain and what hemisphere holds what types of information.

According to Goswami (2004) cognitive neuroscience can offer methods of early detection for those children with special needs. Goswami (2004) believes that much of the research is too specific and that a much wider range of information still needs to be addressed. The history of the brain shows that not all theories were correct and with the advancement of technology, many of the theories that we now believe to be true may in fact not be so accurate in the future (Bergen & Coscia, 2001).

The 1990s were declared the decade of the brain by President Bush (Slegers, 1997). The new theories that were developed in the 1970s about the two hemispheres of the brain that caused scientists to term people as left-brained or right-brained was no longer used (Saunders & Vawdrey, 2002). The knowledge gained from the 1990s has helped educators to understand the simplistic functions of the brain to provide students with strategies that make them learners for life.

THE LANGUAGE CENTER

The breakthrough of technology has helped to make it possible to visualize each part of the brain and what it does. The language center is located in a small area of the brain called the perisylvian region (Eliot, 2001). The perisylvian region surrounds a fissure known as the Sylvian fissure that separates the temporal lobe from the parietal and frontal lobe. These parietal and frontal lobes are located in the left hemisphere where the understandings of speech sounds are found (Eliot, 2001).

Montanaro (2001) states that there are two different stages involved in language acquisition: prelanguage that begins before birth and lasts until the age of 10 or 12 months, and the linguistic stage from the ages of 12 to 36 months. Babies have very little observable external hearing but are actually taking in everything that they hear and the information is hidden inside of them. Montanaro (2001) calls this period "silence" (p. 2). The window for opportunity for language is from birth until age 10 (Slegers, 1997). By the time a child is three years old 97% of children are able to connect 2–3 words to form phrases and simple sentences (Slegers, 1997). Montanaro (2001) states that this period is a "sensitive period for naming things" (p. 2). It is important for the child to learn the correct terminology of words. Baby talk will confuse the child in the long run and the child's vocabulary will not grow.

Language is broken into two categories: words and grammar. These two components are developed at different times and in different areas of the brain. In Eliot's (2001) research she determined that people are able to speak seven hundred speech sounds in one minute of normal speech. It is during this development that the plasticity of the brain continues. This learning window before the age of ten could be an excellent opportunity to teach the child a second language since the brain is already wired for language acquisition.

The brain when viewed with a fMRI shows that children who have normal language skills have lopsided brains (Slegers, 1997). That being the case, it is not surprising that the children with language disorders have brain sides of equal size. Children are born with equal sides of the brain (Slegers, 1997). The right side develops first and grows faster. This is the side of the brain that deals with emotion. The left side of the brain starts to grow later and is in charge of new learning (Slegers, 1997). Both sides of the brain can work independently and can work together.

Siegler (1978) believed that children's spoken language is based on the preoperational period that tends to be more representational rather than transformational. Even though children pass through different stages, it is their environment and specific experiences that affect each child's development. The child's short-term memory regulates the speed at which the child's sequencing progresses. "Memory grows from one unit at age one to two to five units at age four to five years" (Siegler, 1978, p. 54). The predetermined theory is a theoretical based position that language acquisition is innate. The human brain is believed to be wired to master any language. This is the position that brain based researchers have come to believe and have proved to be correct through their many studies and research (Sousa, 2006).

CRITICAL OPPORTUNITIES FOR MAXIMUM LEARNING

The use of technology and research about brain development and its functioning has allowed for researchers to determine when critical opportunities for learning to take place within the different areas of the brain.

One such example of the importance of the windows of opportunity occurred in a preschool that involved a young child adopted from China. This is about a little girl who was adopted from China at age two. Generally the children, mostly girls, are well cared for in the orphanages. Many of these girls are adopted outside of China to parents who wish to have a child. A set of parents from the United States are awaiting the arrival of their precious little girl from China. For bureaucratic reasons, the paperwork to adopt her is tied up in China. China decides to put adoptions on hold for a period of time. The parents in the United States will have to wait longer than they had expected for their child.

Fast forward to the next year, as almost a year goes by; the little girl continues to live at the orphanage. The women that work in the orphanage do not communicate or interact much with the little girl since soon she will be going to the United States and will have a family who will dote on her where she will learn to speak English, which her caregivers do not. The women pay more attention to the others who are not scheduled to be adopted. By the time the waiting period is lifted, the little girl is just shy of her second birthday. The new parents bring home their new daughter to discover that she is unable to communicate with them. At first, it is understandable that there is a communication difficulty due to the fact that the child speaks Chinese.

The parents had learned basic Chinese to make this transition less difficult for her when she arrived in the United States. The child is not able to respond to their Chinese. The parents take her to a therapist that specializes in communication disorders and discovers that the little girl cannot speak Chinese. She is able to communicate with facial expressions and pointing, but she has no words. They try intensive therapy to teach her a few words to communicate. The little girl makes progress but only with her receptive language skills. She is taught sign language to communicate her needs. She is able to sign simple requests, but she still has no words. Why did this happen? What could be wrong?

The doctor determines that this little girl was not talked to as an infant in any language. Her lack of communication caused many of her brain synapses to wither away. Since they were not being used, they could not be developed during the critical window of opportunity. Even though initial language learning continues until age twelve, the child had not even the basic of information given to her for her synapses to develop properly. She is able to communicate with the use of a language board and sign language. As she continues to develop, her communication skills continue to flourish with the intensive language therapy she receives but she is still unable to communicate through spoken language.

This scenario is not typical, but a severe case of when a child is not stimulated with language at birth and does not have interaction with stimuli that would help to increase the windows of opportunity for optimal learning. This example of what happened to this child was written to help explain the importance of building synapses through environment and experiences from birth and what could happen in the absence of stimuli and not using the windows of opportunity for critical learning to take place.

NEW INSIGHTS FOR EDUCATORS

More knowledge about critical windows of learning will help educators develop timelines that concentrate on these windows. Brain research has led to the betterment of early childhood education. Many of the classic theories are still prevalent in today's classrooms. Learning is an individual process and many times it is constructivist in nature. The constructivist theories of Piaget and Vygotsky can be found in the foundations of many early childhood curricula today. Piaget's ages and stages theory of development can be identified with the cognitive growth of the brain. Many other programs and curricula were developed based on theories of the individuals who were popular at the time. For example the theories of Hunt, Skinner, and Piaget helped to bring about Head Start and High/Scope models. Gardner's theory of multiple intelligence is an example of brain based education that promotes whole language learning with the coordination of themes and units (Slegers, 1997).

What does the brain research mean to early childhood education? The advancement of brain research can help to identify at-risk students or children with special needs at a much earlier age than before. Having little or no knowledge of the way the brain functions in relation to learning is not beneficial for the education of the students. New neuroscience discoveries have now been found to disprove old assumptions of how students learn. It is up to the educator to incorporate some of these strategies into their teaching to make the most of the educational experience for the students. Educators can face problems with students' learning due to using old methods of teaching and not using brain research to target these windows of learning.

Learning is different for each child and the types of learning vary based on the age and stage of development the child is in. Until the age of five, children use the right hemisphere for almost all learning. Once the child has reached kindergarten, he is expected to learn in a different manner. Before age five, children learn through exploration and play, after age five, children are expected to sit still and learn at a desk or table. Sequential knowledge is harder for the brain to process. Nonlinear learning in bits and pieces is easier for the brain to process. Each side of the brain processes differently. When the brain is working as a whole, great potential can be achieved (Slegers, 1997). Teachers need to be aware of these processes of the brain to plan curriculum so that it best meets the needs of the children.

CONCLUSION

Each year more and more technological advancements determine new findings about the brain. Have scientists and educators found out all there is to know about the brain and how it functions? No, time will tell as to what new things will be learned about the brain and its functioning. Several years from now educators may find that some of the theories that were in place were not found to be correct or need to be improved upon. The brain is multifaceted, there is still so much more to learn and discover.

As the century continues, more brain research will bring about new and improved information that can be used to make education and learning better. Is it possible that we can even be able to pinpoint difficulties to the point of being able to fix them and see a decrease of disabilities in students? One cannot know the range of knowledge gained from studies and research but one would like to believe that it could be possible.

REFERENCES

Bergen, D., & Coscia, J. (2001). *Brain research and childhood education: Implications for educators.* Olney, MD: Association for Childhood Education International.

Eliot, L. (2001). Language and the developing brain. *NAMTA Journal*, 26(2), 8–60.

Goswami, U. (2004). Neuroscience and education. *British Journal of Educational Psychology*, 74(1), 1–14.

Montanaro, S. (2001). Language acquisition. *NAMTA Journal*, 26(2), 1–7.

Saunders, A. D., & Vawdrey, C. (2002). Merging brain research with educational learning principles. *Business Education Forum*, 1, 44–46.

Siegler, R. S. (2000). *Childhood cognitive development: The essential readings.* Malden, MA: Blackwell.

Siegler, R. S. (1978). *Children's thinking: What develops?* New York: Halsted Press.

Slegers, B. (1997). Brain development and its relationship to early childhood education. *Presented at EDEL seminar in elementary education,* Long Beach, CA, April 17, 1997.

Sousa, D. A. (2006). *How the brain learns* (3rd ed.). Thousand Oaks, CA: Corwin Press.

Brains Show Two Sides of Language Function*

By Bruce Bower
Science News, June 15, 2002

Damage to the brain's left side often undermines language abilities. Occasionally, so does right-brain damage. Still, a lucky few individuals can suffer injury to either side and retain their verbal skills.

Thanks to a device that temporarily blocks activity in specific brain areas, scientists have uncovered a likely explanation for this linguistic reversal of fortune. Some individuals, conclude neurologist Stefan Knecht of the University of Münster in Germany and his colleagues, have strong language capabilities in both halves of their brains. These individuals have enough neural leverage to withstand a block on one side or the other, the researchers report.

In contrast, the majority of people, whose language capability relies primarily on left-brain structures, exhibit temporary verbal losses during brief disruptions of those areas. Comparable difficulties occur when the right brain is blocked in the smaller proportion of people who depend on that side to coordinate language use, the scientists report in an upcoming Nature Neuroscience.

"Some people possess a network of language areas on both sides of the brain that resists localized damage," Knecht says.

Knecht's group used functional magnetic resonance imaging to measure blood-flow changes in the brains of 324 men and women as they thought about as many words as possible beginning with letters shown on a computer screen. This method indirectly gauges cells' activity throughout the brain.

The results suggest that about 1 in 10 people exhibits two-sided neural organization. Another 1 in 10 displays right-brain coordination of language. Both these brain patterns occur mainly in left-handers.

No difference in intelligence, creativity, or academic achievement shows up among the groups with left-brain, right-brain, or two-sided language control.

The researchers further studied 20 individuals. Each of these volunteers performed a second language task in which he or she noted as quickly as possible whether animal pictures were correctly labeled. These participants also wore caps with coils that briefly generated magnetic fields. This process, known as transcranial magnetic stimulation (SN: 9/23/00, p. 204), enabled the scientists to suppress cell activity in either left- or right-brain tissue.

The six people identified as having two-sided language activity continued to do well on the picture-word task during both left- and right-brain suppression. However, performance plummeted for the seven participants with left-brain specialization for language and the seven with right-brain specialization when parts of their crucial neural hemisphere were suppressed.

Curiously, these individuals did slightly better at matching pictures and words during brief disturbances of their other hemisphere. Since the nondominant hemisphere analyzes the context of speech, its suppression might make it easier to perform simple word tasks, Knecht theorizes.

The new report confirms prior suspicions about left- and right-brain language functions, remarks psychologist Glyn W. Humphreys of the University of Birmingham in England. "The nature of language representation in the brain still remains unclear," he adds.

5

The Brain and Aging

Editor's Introduction

New evidence regarding the brain's plasticity, or ability to change and grow new neurons over time, is good news for anyone concerned with aging. It suggests that middle-aged and elderly people have the power to combat the kind of age-related mental deterioration once thought to be inevitable. While there is reason to be hopeful, the fact remains that aging brains undergo many changes. They shrink, for one thing, and by the time people celebrate their 60th birthdays, their brains start to lose between .5 and 1 percent of their volume each year. What's more, molecules known as free radicals—the natural by-products of metabolism—begin to degrade the myelin coatings that insulate neurons. This leads to a slowdown in mental processing, making it harder for seniors to recall pieces of information or perform multiple tasks simultaneously. "You may have the same size hard drive," Dr. Howard Fillit, executive director of the Institute for the Study of Aging, told Sharon Begley for *Newsweek*, "but the processing speed of your computer is slowing down."

More troubling is Alzheimer's Disease, a condition that kills brain cells and causes dementia. Alzheimer's seems to affect anyone who lives long enough to contract it, and according to estimates, 39 percent of people aged 90 to 95 suffer from the disease. Scientists remain unsure of its mechanisms, though many believe cell destruction is caused by a protein called amyloid-beta, or A-beta, which collects and forms insoluble plaques. As of 2008, 4.5 million Americans were afflicted with Alzheimer's, and by 2050 that number is expected to reach 16 million. While researchers have yet to find a cure, many are urging seniors to keep their minds limber by learning new skills and maintaining social relationships. Exercise has also been shown to enhance cognitive ability, perhaps because it prevents "microstrokes" in the brain's blood vessels.

The articles in the first chapter provide an overview of how the brain changes with age. In the first entry, "The Disappearing Mind," Geoffrey Cowley, Anne Underwood, and Andrew Murr examine various methods of diagnosing and treating Alzheimer's Disease. Though the search for a cure has thus far proved fruitless, the authors remain hopeful, citing promising pharmaceutical research. In the next entry, "The Brain in Winter," Begley focuses on how the brain shrinks and slows down over time. As the speed with which the brain transmits electrical impulses starts to decrease—something that happens around the age of 50—people take longer to remember things. Age also eliminates dopamine receptors, areas that activate pleasure and control impulses.

Begley accentuates the positive with "New Research Finds Some Brain Functions Actually Improve With Age," the next selection in the chapter. Citing recent findings, she suggests that seniors enjoy enhanced "semantic memory," the ability to recall facts and figures. Elderly people also benefit from "cognitive templates," or memorized solutions to problems they encountered in the past. It's for this reason that some veteran professionals perform better than their young counterparts. The chapter continues with "The Outlook for Alzheimer's Disease," in which Kimbal E. Cooper and Tyler A. Kokjohn highlight the likely causes of and current treatments for the debilitating brain disease. The authors end with the question of whether Alzheimer's might be an evolutionary tool designed to bring about death. In "Neuron Killers," the next article, Tina Hesman Saey approaches Alzheimer's like a detective, running down the list of proteins that scientists believe could be responsible for killing brain cells.

In the final selection, "Keeping Your Brain Fit," Christine Larson discusses how brain-training programs, exercise, good eating habits, and regular social interaction may help seniors stave off negative effects of aging, such as dementia.

The Disappearing Mind[*]

By Geoffrey Cowley
Newsweek, June 24, 2002

Do senior moments scare you? Eight years ago Nancy Levitt had one that would unsettle anyone. She was in her mid-40s at the time, and watching her father drift into the late stages of Alzheimer's disease (several aunts and uncles had suffered similar fates). Levitt's son was about to graduate from high school, so she called a mail-order company to order fluorescent-light sticks for him and his friends to wear around their necks at a party. When her package arrived in the next day's mail, the receipt and postmark knocked her flat. She had ordered the same gift a few days earlier—and lost all recollection of it. "I just freaked," she says. Suddenly, every forgotten name, misplaced pencil and misspelled word became a prophecy of doom. Was she getting Alzheimer's herself?

When Levitt sought testing at UCLA, researchers gave her the usual cognitive tests—name some simple objects, repeat a list of words—and assured her she was fine. But the occasional lapses continued, so she returned to the same clinic several years later and enrolled in a study aimed at distinguishing early Alzheimer's from run-of-the-mill forgetfulness. This time the researchers didn't just talk to her. They placed her under a scanner and recorded detailed images of her brain, both at work and at rest. Alzheimer's disease has traditionally been diagnosed by exclusion. If you lagged significantly on a memory test—and your troubles couldn't be blamed on strokes, tumors or drug toxicity—you were given a tentative diagnosis and sent on your way. To find out for sure, you had to die and have your brain dissected by a pathologist. Levitt didn't have to do any of that. By looking at the images on his video screen, Dr. Gary Small was able to give her some reassuring news. She didn't have Alzheimer's disease—and the odds were less than 5 percent that she would develop it any time soon. Levitt calls the images "the most wonderful thing I've ever seen."

Technology is changing all of medicine, but it is positively transforming our understanding of Alzheimer's. Armed with state-of-the-art PET scanners and MRI machines, specialists are learning to spot and track the disease in people who have yet to suffer symptoms. It's one thing to chronicle the brain's disintegration, quite another to stop it, but many experts are predicting success on both fronts. Drugmakers now have two dozen treatments in development. And unlike today's medications, which offer only a brief respite from symptoms, many of the new ones are intended to stall progression of the disease. As Alzheimer's runs its decades-long course, it replaces the brain's exquisite circuitry with mounds of sticky plaque and expanses of dead, twisted neurons. No drug will repair that kind of damage. But if the new treatments work as anticipated, they'll enable us to stop or slow the destruction while our minds are still intact. A decade from now, says Dr. Dennis Selkoe of Harvard Medical School and Boston's Brigham and Women's Hospital, physicians may monitor our brain health as closely as our cholesterol levels—and stave off Alzheimer's with a wave of the prescription pad.

Until we can control this awful illness, early detection may seem a fool's errand. "With diagnostics ahead of therapeutics, there's a lot of potential for harm," says University of Pennsylvania ethicist Arthur Caplan. He worries that entrepreneurs will peddle testing without counseling, leaving patients devastated by the findings. He wonders, too, whether employers and insurers will abandon people whose scans show signs of trouble. Advocates counter that early detection can help patients make the most of today's treatments while giving them time to adjust their plans and expectations. With so many people at risk, they say, anything is better than nothing. Some 4 million Americans have Alzheimer's today, but the number could hit 14 million by 2050 as the elderly population expands.

The diagnostic revolution began during the 1990s, as researchers learned to monitor neurons with an imaging technique called PET, or positron-emission tomography. Unlike an X-ray or CT imaging, PET records brain activity by homing in on the glucose that fuels it. And as Small's team has discovered, it can spot significant pathology in people who are still functioning normally. Instead of glowing with activity, the middle sections of their brains appear dim and torpid. And because Alzheimer's is progressive, abnormal scans tend to become more so with time. In a study published last fall, UCLA researchers scanned 284 people who had suffered only minor memory problems. The images predicted, with 95 percent accuracy, which people would experience dementia within three and a half years.

PET scanning has yet to transform patient care; few clinics have the machines, and Medicare doesn't cover their use. But scientists are now using the technique to see whether drugs already on the market (such as the anti-inflammatory ibuprofen) can slow the brain's decline. And PET is just one of several potential strategies for tracking preclinical Alzheimer's. San Die-go researchers have found that seniors who score inconsistently on different mental tests are at increased risk of dementia—even if their scores are generally high. And in a study published this spring, researchers at the Oregon Health Sciences University hit upon three signs of imminent decline in octogenarians. The 108 participants were all healthy at the start of the study, but nearly half were demented six years later. As it turned out, they had entered the study with certain traits in common. They walked more slowly than their peers, requiring nearly two extra seconds for a 30-foot stroll. They lagged slightly on memory tests. And their MRI scans revealed a slight shrinkage of the hippocampus, a small, seahorse-shaped brain structure that is critical to memory processing. The changes were subtle, says Dr. Jeffrey Kaye, the neurologist who directed the study, but they presaged changes that were catastrophic.

Powerful as they are, today's tests show only that the brain is losing steam. The ideal test would reveal the underlying pathology, letting a specialist determine how much healthy tissue has been replaced by the plaques and tangles of Alzheimer's. It's not hard to fashion a molecule that will highlight the wreckage. Unfortunately, it's almost impossible to get such a probe through the ultrafine screen that separates the brain from the bloodstream. If a probe is complex enough to pick out plaques and tangles, chances are it's too large to pass from the bloodstream into the brain. At UCLA and the University of Pittsburgh, researchers have developed probes that are small enough to get through, yet selective enough to provide at least a rough measure of a person's plaque burden. At Brigham and Women's Hospital, meanwhile, radiologist Ferenc Jolesz is trying to open the barrier to bigger, better probes. His technique employs tiny lipid bubbles that gather at the gateway to the brain when injected into the bloodstream. The bubbles burst when zapped with ultrasound, loosening the mesh of that ultrafine screen and allowing the amyloid probe to enter. Lab tests suggest the screen will repair itself within a day, but no one yet knows whether it's safe to leave it open that long.

One way or another, many of us now seem destined to learn we have Alzheimer's disease while we're still of sound mind. The question is whether we'll be able to do anything more constructive than setting our affairs in order and taking a drug like Aricept to ease the early symptoms. Fortunately the possibilities for therapy are changing almost as fast as the diagnostic arts. Experts now think of Alzheimer's not as a sudden calamity but as a decades-long process involving at least a half-dozen steps—each of which provides a target for intervention. Slowing the disease may require four or five drugs rather than one. But as AIDS specialists have shown, the right combination can sometimes turn a killer into a mere menace.

Though experts still quarrel about the ultimate cause of Alzheimer's, many agree that the trouble starts with a scrap of junk protein called amyloid beta (A-beta for short). Each of us produces the stuff, and small amounts are harmless. But as A-beta builds up in the brain, it sets off a destructive cascade, replacing healthy tissue with the plaques seen in Alzheimer's sufferers. No one knew where this pesky filament came from until 1987, when researchers discovered it was part of a larger molecule they dubbed the amyloid-precursor protein (APP). Thanks to more recent discoveries, they now know exactly how the parent molecule spawns its malevolent offspring.

APP is a normal protein that hangs from a neuron's outer membrane like a worm with its head in an apple. While performing its duties in and around the cell, it gets chopped up by enzymes called secretases, leaving residues that dissolve in the brain's watery recesses. Occasionally, however, a pair of enzymes called beta and gamma secretase cleave APP in just the wrong places, leaving behind an insoluble A-beta fragment. Some people produce these junk proteins faster than others, but after seven or eight decades of service, even the healthiest brain carries an amyloid burden. When it reaches a certain threshold, the brain can no longer function. That's why Alzheimer's dementia is so rampant among the elderly. Given enough time, anyone would develop it.

The ideal Alzheimer's remedy would simply slow the production of A-beta— by disabling the enzymes that fabricate it. Elan Corp. was the first drugmaker to try this tack. During the mid-'90s its scientists developed several gamma-secretase blockers and tested them in animals—only to find that they sometimes derailed normal cell development, damaging bone marrow and digestive tissues. A few companies are still pursuing gamma blockers, but beta secretase now looks like a safer target for therapy. More than a half-dozen drugmakers are now working on beta inhibitors. "In the industry," says Dr. Ivan Lieberburg of Elan, "we're hoping that the beta-secretase inhibitors will have as much therapeutic potential as the statins." Those, of course, are the cholesterol-lowering medicines for which 35 million Americans are now candidates.

Secretase inhibitors may be our best hope of warding off Alzheimer's, but they're not the only hope. As scientists learn more about the behavior of A-beta, they're seeing opportunities to disarm it before it causes harm. One thing that makes A-beta fragments dangerous is their tendency to bind with one another to form tough, stringy fibrils, which then stick together to create still larger masses. Three companies are now testing compounds designed to keep A-beta from forming fibrils—and at least two other firms are working to keep fibrils from aggregating to create plaque. All of their experimental drugs have helped reduce amyloid buildup in plaque-prone mice, suggesting they might help people as well. But human studies are just now getting underway.

Suppose for a moment that all these strategies fail, and that amyloid buildup is simply part of the human condition. As Selkoe likes to say, there's more than one way to keep a bathtub from overflowing. If you can't turn down the faucet, you can always try opening the drain. Recognizing that most of the people now threatened by Alzheimer's have already spent their lives under open amyloid faucets, researchers are pursuing several strategies for clearing deposits from the brain. One elegant idea is to mobilize the immune system. Three years ago Elan wowed the world by showing that animals given an anti-amyloid vaccine mounted fierce attacks on their plaques. Vaccinated mice reduced their amyloid burdens by an astounding 96 percent in just three months. The vaccine proved toxic in people, triggering attacks on normal tissue as well as plaque, but the dream isn't dead. Both Elan and Eli Lilly are now developing ready-made antibodies that, if successful, will target amyloid for removal from the brain without triggering broader attacks by the immune system.

Even later interventions may be possible. As a person's amyloid burden rises, so does the concentration of glutamate in the brain. This neurotransmitter helps lock in memories when it's released in short bursts, but it kills neurons when chronically elevated. At least two teams are now betting they can rescue cells surrounded by amyloid, simply by shielding them from glutamate. One possible life jacket is a drug called Memantine, which is already approved in Europe. It covers a receptor that lets glutamate flow freely into neurons, but without blocking the glutamate bursts needed for learning and memory. New York's Forest Laboratories is now launching an American trial of the drug, and hoping for approval by next year.

If even half these treatments fulfill their promise, old age may prove more pleasant than today's projections suggest. For now, the best we can expect is an early warning and perhaps a year or two of symptomatic relief. That may seem a paltry offering, but it's a far cry from nothing. As Small argues in a forthcoming book called "The Memory Bible," people at early stages of Alzheimer's can do a lot to improve their lives, but few of them get the chance. Three out of four are already past the "moderate" stage by the time their conditions are recognized. Some may find solace in ignorance. But the case for vigilance is getting stronger every day.

The Brain in Winter[*]

By Sharon Begley
Newsweek, September 1, 2001

Let's clear up one thing right away. The one "fact" everyone knows about the aging brain is that it loses something like 10,000 neurons a day starting (in the optimistic version) at the age of 65 or (in the might-as-well-give-up-now version) at 30. Even in a brain with 100 billion neurons, losing 10,000 of those babies every day (3.65 million a year!) sounds pretty dire. But while cell loss in the aging brain has been neurological dogma for decades, it's flat-out wrong: thanks to improved techniques for counting cells, researchers have shown that any loss of brain cells is minimal and has few, if any, real-world consequences. Even better is the overthrow of that other textbook tenet, that you're born with all the brain neurons you'll ever have. "We are not losing cells in significant numbers, and the older brain is capable of generating new cells," says neuroscientist Molly Wagster of the National Institute on Aging. Those discoveries put a decidedly different spin on the graying of our gray matter. "If brains do not lose cells in the numbers once supposed, and if some surviving cells retain a capacity to replicate," says molecular biologist Lawrence Whalley of the University of Aberdeen in Scotland, "there may be real prospects of slowing or even preventing some of the worst effects of brain aging."

Which is to say—and there's no getting around this—there are indeed some clear effects of age on the brain. It is subject to the same basic biology as the rest of the body, from built-in senescence to normal wear and tear to damage by free radicals (no, not roving activists, but rogue molecules produced during normal metabolism). As a result, the brain undergoes one quite dramatic change as the decades pile up. Starting at about age 50, it shrinks. The average 50-year-old's brain weighs three pounds; 15 years later it weighs 2.6 pounds. Most of the shrinkage comes from brain cells' losing water content. "Gaps between the folds of the cortex widen and the large spaces inside the brain enlarge," Whalley says.

The frontal lobes, seat of higher thought, show more thinning than any other neighborhood. They can lose an average 0.55 percent of their volume every year after 50 or so (twice the rate of loss in other regions), with the result that they can shrink 30 percent between the ages of 50 and 90, says neuroscientist Mark Mattson of the National Institute on Aging. The frontal lobes perform so many complex functions that scientists are only beginning to sort out the effect of this shrinkage, but "there is now a consensus that the psychological changes of aging"—such as, perhaps, in social judgment and emotional responsiveness—"reflect deterioration of the frontal lobes," says Whalley. So might impaired attention, impulse control and difficulty focusing on several things at once.

Farther back in the brain, the hippocampus can lose about 20 percent of its volume between the ages of 50 and 90. This little sea-horse-shaped structure is crucial to forming as well as retrieving memories. In normal aging, its levels of the neurotransmitter acetylcholine fall. Since acetylcholine is one of the molecules by which neurons communicate, its scarcity in the hippocampus provides "one of the most likely explanations of memory deficits in otherwise healthy old people," says Whalley. And in not-healthy older people? There is no definitive way to diagnose Alzheimer's (except through autopsy), but a test called the Mini-Mental State Exam comes close. It asks the date or year, for example, and has the person count backward from 100 by sevens, recall three objects named a few minutes before, read and obey the sentence "close your eyes," as well as do other simple tasks. A low score, combined with confusion about places, a loss of initiative and problems recognizing friends are more likely to presage Alzheimer's than forgetting an appointment.

Although brain shrinkage mostly reflects loss of water, it's also the result of shrinking dendrites, says NIA's Mattson. Dendrites are the scraggly fibers that grow out of the cell body of a neuron like snakes from Medusa's head. Their job in life is to receive the signal carried by an adjoining neuron's axon, the long, thin fiber that extends from the cell body like a string on a balloon. Axons carry electrical impulses. The axon from one neuron connects, typically, to the dendrites of another, across a gap called a synapse. When synapses are strong, transmission is more likely; the formation of strong synapses underlies learning and remembering. When dendrites are numerous, there's a greater chance of neuron-to-neuron transmission's continuing. It is that transmission that constitutes the great neural symphony that lets us think, feel, remember and dream. When dendrites thin out, there is a greater chance of neuronal transmission's coming to a halt, something that may underlie some of the difficulties that many elderly people have with, for instance, following grammatically complex sentences and abstruse logic.

If you find that tracking two conversations simultaneously is impossible or that instructions for connecting your new printer are indecipherable, blame what cognitive scientist Denise Park of the University of Michigan calls "the most reliable finding in the field" of brain aging: that older neurons process signals more slowly. "You may have the same size hard drive," says Dr. Howard Fillit, executive director of the New York-based Institute for the Study of Aging, "but the processing speed of your computer is slowing down." Reaction time slows. The ability to retrieve old information and learn new can be impaired. Multitasking can be more of a challenge than it is for your IM'ing/CD-listening/homework-doing grand-kid: the older brain is less tolerant of interference and distractions. If recalling an obscure fact requires searching many memory nodes—like following multiple hypertext links—then you'll notice the slower processing. Complicated tasks that require pulling together bunches of scattered memories may take longer and re-quire more effort. "Slowing explains almost all the age-related changes in cogni-tive abilities," says Park.

The slowdown probably reflects several changes in the brain. Basic metabolism generates those nasty free radicals. Giving promiscuity a bad name, these mol-ecules combine and react with just about any other molecules they sidle up to, including myelin, the fatty sheath that insulates neurons. When free radicals react with fatty acids in the myelin, the myelin degrades. The velocity with which elec-trical signals zip around the brain falls. "A reduced conduction velocity of axons would slow down information processing in the brain," says Mattson. "This prob-ably occurs in all brains from middle age onward," says Whalley, whose book "The Aging Brain" will be published in November. "Since both mental slowing and the structural changes in myelin are universal in aging, they are probably linked."

Free radicals can also thin out dendrites like Edward Scissorhands on a manic day. The result is that a signal originally destined to jump from the transmitting axon to the receiving dendrites no longer has any place to go. With age, the num-ber of dendrites in a given area—their density—decreases, leaving the brain less connected. As a result, brain signals "travel along more circuitous pathways, which results in slowing," says Michigan's Park. "It's like a message over the Internet that goes through too many nodes." The result is that slight hesitation as you search for the name of that movie . . . you know, the one where Audrey Hepburn is the chauffeur's daughter . . . But drawing a blank is not a sign of Alzheimer's. That disease is marked not by something as subtle as thinning dendrites. Instead, Alzheimer's brains are full of wads of sticky proteins called amyloid plaques that displace and sometimes kill neurons, and twisted filaments called neurofibrillary tangles that can make neurons wither away.

Park's Internet comparison is apt because, like the Web, what matters in the brain is connections—synapses. Usually, synapses get stronger with use. That's why if you practice and practice a language, an instrument or an athletic move, you have a better chance of mastering it: you literally wire into your brain the req-uisite neuronal connection. With age, those connections begin to change. "There seem to be decreases in synaptic density in certain regions of the brain, like the hippocampus, that are important for thinking," says neurologist Marilyn Albert of Harvard University. "There is not much nerve-cell loss, but there is some loss of connectivity."

Newly formed synapses also seem to lose strength more quickly in old brains. Neurologist Carol Barnes of the University of Arizona had been interested in memory and aging since she was a graduate student in the 1970s and her grandfather was beginning to get lost during long walks. Barnes devised an experiment that might show why. She put young and old rats in mazes, luring them to the end by a pile of chocolate sprinkles. As each animal solved the maze, Barnes monitored the electrical activity in its hippocampus, the seat of spatial memories. An hour or so later, she dropped each animal back in. Young rats activated the same "hippocampal map" that got them through the twists and turns before. So did most older rats. But one third of the geezer rats lost their way and tried a different route, as hippocampal activity changed completely: they were apparently unable to retrieve the hippocampal map recorded in their first run. Confusion about places can be a sign of early Alzheimer's.

"Because new synapses are not as strong in older rats, the maps they create are less [stable]," says Barnes. As a result they—and her peripatetic grandfather—have to keep learning new routes, and possibly other new information, over and over. "You can teach an old dog new tricks," says Barnes. "But it may take longer, and the results may fade faster." Annoying, sure, but it may reflect a shrewd strategy on nature's part. If old, established connections underlie walking, talking, reading, writing, dressing and other basics, then perhaps making it difficult for new connections to override them is a good idea.

If life seems to lose some of its zest with each passing year, there may be a reason beyond intimations of mortality. Brain neurons are studded with receptors, molecules that dangle into synapses to snare passing neurotransmitters like a fishhook snaring a trout. When caught, the neurotransmitter dopamine can trigger neuronal activity associated with pleasure. Age, however, seems to eliminate some dopamine receptors, according to a 2000 paper by Nora Volkow of Brookhaven National Laboratory and colleagues. With each decade starting at the age of 20, 6 percent of dopamine receptors called D2 disappear. Thus a once pleasing sight or a once joyous memory might not inspire the same level of pleasure it once did, when dopamine receptors were more numerous.

But there's more. When dopamine receptors decrease, so does brain activity in regions associated not with pleasure and enjoyment but with cognition, according to PET scans the scientists took of their healthy volunteers. Metabolism, finds Volkow's group, decreases with age in both the frontal regions, the site of problem solving, abstract thinking and multitasking, as well as the anterior cingulate gyrus, responsible for attention span and impulse control. So while Grandpa's habit of blurting out whatever's on his mind might reflect the impatience with social niceties that comes with mature wisdom, it might also result from a general slowdown in the anterior cingulate gyrus.

A slower metabolism is not the only chemical warfare directed at the brain in winter. Aging disrupts one of its most finely tuned systems, the "negative feedback" by which high levels of the stress hormone cortisol trigger a production shutdown and low levels ramp production up. This cortisol system seems to break down with age, with the result that the brain gets bathed in the stuff. "Cortisol can influence many aspects of brain function," says Whalley, including learning and biological rhythms—mostly for the worse. People over 70 spend much less time in deep sleep and more time in light sleep than younger people do (which is why the slightest noise can rouse a sleeping 75-year-old); although the causes of poor sleep in the elderly aren't known for sure, messed-up biorhythms can't help. Chronic stress can also reduce axon sprouting and dendrite branching, both of which underlie learning and memory. Not everyone over 50 starts churning out cortisol, but scientists at last year's workshop on aging and cognition concluded that anything that reduces stress can only help the brain.

The biggest recent surprise in neuroscience is the finding that as the brain ages it creates new neurons. In 1998, neuroscientists reported that they had examined the brains of five cancer patients who had just died. Contrary to all expectations, the scientists found evidence of living, dividing precursor cells giving birth to somewhere between several hundred and 1,000 new cells every day. This discovery sent a shock through neuroscience. The decades-long dogma that the brain creates no new neurons after infancy had become so entrenched that researchers who argued otherwise (and even glimpsed evidence) were ridiculed and even ostracized. Although scientists do not yet know what good the perpetual supply of neurons does, every little bit surely helps. Maybe the new recruits replace cells lost to stress hormones or free radicals; maybe they weave themselves into the synapses that support new memories.

The discovery of neurogenesis reinforces the emerging notion of "how little brain deterioration there is in a healthy individual," says Dr. Jeffrey Cummings of UCLA. Many of the calamities or just plain nuisances previously blamed on aging actually reflect illness, and nowhere is that truer than in the aging brain. If the rest of the body is aging badly, so will the brain. "In normal aging," says Cummings, "there is surprisingly good brain function. Because the brain is the instrument of survival, it has been spared as much as possible from insults that occur to the rest of the body." Of course, with your mental faculties almost as sharp as ever, you're more likely to notice those other insults.

The Upside of Aging[*]

New Research Finds Some Brain Functions Actually Improve with Age

By Sharon Begley
Wall Street Journal, February 16, 2007

The aging brain is subject to a dreary litany of changes. It shrinks, Swiss cheese-like holes grow, connections between neurons become sparser, blood flow and oxygen supply fall. That leads to trouble with short-term memory and rapidly switching attention, among other problems. And that's in a healthy brain.

But it's not all doom and gloom. An emerging body of research shows that a surprising array of mental functions hold up well into old age, while others actually get better. Vocabulary improves, as do other verbal abilities such as facility with synonyms and antonyms. Older brains are packed with more so-called expert knowledge—information relevant to your occupation or hobby. (Older bridge enthusiasts have at their mental beck-and-call many more bids and responses.) They also store more "cognitive templates," or mental outlines of generic problems and solutions that can be tapped when confronting new problems.

Eric Kandel, 77 years old, who shared the 2000 Nobel Prize in medicine, maintains an active lab at Columbia University and mentors younger scientists. "I think I do science better than I did when I was younger," he says. "In science, judgment is so important, and I now have a better understanding of which problems are important and which aren't."

Growing awareness that old brains aren't necessarily senile brains is already fueling a slew of consumer offerings. Brain exercises developed for older adults by Posit Science Corp. in San Francisco are being offered by retirement communities, senior centers and assisted-living facilities, as well as by insurers such as Humana to their Medicare enrollees. The computer-based program includes exercises intended to improve memory and attention, as well as sharpness of hearing. Continuing, peer-reviewed studies conducted by Posit scientists suggest it can roll back the mental agility calendar by at least a decade.

Some retirement communities and assisted-living centers are installing a touch-screen-based cognitive fitness program developed by Dakim Inc. of Santa Monica, Calif., that gives seniors practice on seven cognitive skills, including language and the kind of visual-spatial processing that helps you read a map. The system uses "age-appropriate" film and audio clips, such as Jimmy Stewart movies, as the basis for short-term memory exercises and adds new exercises.

CHANGING PUBLIC POLICY

Discoveries of brain functions that hold up, or even improve, through the decades could affect corporate and public policy. As baby boomers age, many are resisting mandatory retirement. In January, a special committee of the New York State Bar Association recommended that law firms abandon the practice. Air-traffic controllers are asking federal agencies to reconsider the requirement that they retire at age 55, and the Federal Aviation Administration in January proposed pushing back the mandatory retirement age for commercial pilots, which is currently 60.

The emerging neuroscience is on their side. One of the most robust cognitive abilities is semantic memory, which is recollection of facts and figures. "Semantic memory is relatively resistant to the effects of aging," says psychology professor Arthur Kramer of the University of Illinois, Urbana-Champaign. Semantic memory includes vocabulary, which increases with age so reliably (at least in people who continue reading) that a younger person should never challenge a sharp 75-year-old to a crossword puzzle.

Expert knowledge—information about an occupational or even hobbyist specialty—resists the effects of aging, too, which is why mumbling "accrued postretirement liabilities" to an 80-year-old actuary makes his relevant synapses fire as robustly as they did at age 40. Synapses that encode expert knowledge "are written in stone," says neuroscientist John Morrison of the Mount Sinai School of Medicine in New York.

The longevity of expert knowledge and cognitive templates lies behind the finding that air-traffic controllers in their 60s are at least as skilled as those in their 30s. When Prof. Kramer of Illinois and a colleague at the Massachusetts Institute of Technology gave older controllers standard lab tests for reaction speed, memory, attention and the like, they found the usual: Performance declined compared with that of 30-somethings.

But on more fast-paced, complex—and hence realistic—tests in which they juggled multiple airliners and handled emergencies, the senior controllers did as well as or better than the young ones. They kept simulated planes safely away from each other, and when they ordered planes to change their altitude, heading or speed to avoid a collision, they used fewer commands than younger ones. It was as if their experience had equipped them with the most efficient algorithm for keeping the planes safely spaced.

"Their experience and their knowledge of aircraft types and strategies they've used for years can compensate for a decline in these other abilities," says Prof. Kramer, who has submitted the study to a science journal. The findings, he says, suggest the need to revisit "the whole notion of when we need to retire people, since their ability to do these complex tasks resists decline."

That 60-somethings can mentally juggle multiple 747s seems to go against the idea that aging hurts the ability to pay attention. But studies show that selective attention, the ability to focus on something and resist distractions, doesn't decline with age. For controllers, that means they can focus on planes in their sector despite a hubbub of activity in the control tower. For other seniors, it means no problem keeping eyes and mind on a highway despite flashing road signs or noisy passengers.

The biggest benefit of an older brain is that fewer real-life challenges require deliberate, effortful problem-solving. Where once it took hours of methodical scrutiny to understand a prospectus, for instance, older lawyers and investment bankers can zoom in on crucial sections and fit them into what they already know.

Elkhonon Goldberg, a neuropsychologist who has a private practice and is a professor at New York University School of Medicine, finds that he can also grasp the essence of data presented in scientific papers more readily than he once could, something that more than makes up for losses in other mental realms. "I am not nearly as good at laborious, grinding, focused mental computations," he says, "but then again, I do not experience the need to resort to them nearly as often."

While younger brains solve problems step-by-step, older brains call on cognitive templates, those generic outlines of a problem and a solution that worked before. It's the feeling you get when you see that a new situation or problem belongs to a class of situations or problems you have encountered before, with the result that you don't have to attack them methodically. Yes, older people forget little things, and may have occasional attention lapses, but their cognitive templates are so rich that they more than hold their own. Their brains can keep up even with a diminished supply of blood and oxygen.

PROFESSIONAL BENEFITS

As a result, older professionals can readily separate what's important from what's not, a big reason so many of them fire on all cognitive cylinders well past age 65. "I'd say that the ability to make a significant contribution as a lawyer actually increases with time, experience and age," says attorney Mark Zauderer, 60, a partner in the New York law firm Flemming Zulack Williamson Zauderer.

In complex business litigation, he says, where pretrial discovery can yield enough documents to fill a warehouse, "a lawyer must be able to sort the wheat from the chaff, to take all these facts and extract only those that support winning themes. A senior lawyer is in the best position to do that, and to have the courage to discard facts—even those on your side—that will only distract the court or the jury."

"Some things you just need to grind into your system for many years until they become automatic and seemingly effortless," says Naftali Raz of the Institute of Gerontology at Wayne State University in Detroit. "That may be the key. Automatic functions are least sensitive to aging. So, if the decisions are based on knowledge and skill, older folks may have an advantage over younger decision makers just because they have to do less mental heavy lifting."

More research is coming. Although studies on aging have long focused on diseases such as Alzheimer's, scientists are increasingly investigating healthy aging, trying to discover which factors allow some people to resist the usual ravages of time, and to get a better sense of how well older adults can function. The National Institutes of Health, the nation's leading funder of biomedical research, doesn't break out "healthy aging" as a separate budget item, but spokeswoman Linda Joy says that more funding is going to studies of people who reach their 60s, 70s and beyond with little or no disease. Scientists hope that by identifying which mental functions are largely untouched by aging, they will be able to develop treatments or exercises to shore up functions that do deteriorate.

The benefits that come to the mind and brain with age extend beyond thinking. They also include a greater ability to put yourself in another person's mind, empathizing and understanding his thought processes—emotional wisdom. Civil engineer Samuel Florman, 81, remains active in his Scarsdale, N.Y., construction company and says that as he has grown older, he "has gotten better with people, more understanding of young people and more patient with aggressive ones. I'm more savvy about when to rush and when not to."

CONTROLLING ANGER

That likely reflects the older brain's greater control over emotions, especially negative ones such as impatience and anger. A 2006 study of 250 people ranging in age from adolescence to their late 70s documented for the first time "positive changes in the emotional brain," according to the Society for Neuroscience, which publishes the *Journal of Neuroscience*. In the experiment, Leanne Williams of the University of Sydney showed the volunteers pictures of faces expressing emotions. Using fMRI brain imaging, it was found that circuits in "medial prefrontal" areas—right behind the forehead—were more active in older people than younger people when processing negative emotional expressions. The greater activity suggests better control of reactions to other people's anger, fear and the like. This greater sensitivity seems to translate into decreasing neuroticism, and greater emotional equanimity.

That doesn't mean older brains flatline when it comes to sensitivity. Instead, they often show a keen emotional intelligence and ability to judge character. Elderly volunteers given a list of behaviors that describe a made-up person ignored irrelevant information (favorite color, place of birth) when asked to judge the person's character and focused on revealing traits better than younger people did, according to research by Thomas Hess, a professor of psychology at North Carolina State University. They were more likely to infer correctly that the person was dishonest, kind or intelligent—a skill that is arguably more important than the ability to memorize a list of words in a lab experiment.

The Outlook for Alzheimer's Disease[*]

By Tyler A. Kokjohn and Kimbal E. Cooper
The Futurist, September/October 2005

The rise of Alzheimer's disease in recent decades is a tragic side effect of a great success story: the increase in human longevity.

Alzheimer's disease is the relentless destruction of brain tissue, which causes a debilitating loss of mental capacity. Despite intensive study, only a few mitigating treatments exist, and the disease remains wholly incurable. So little is understood regarding its root causes that it is impossible even to provide definitive advice as to how it may be avoided. But one unsettling fact is clear: Alzheimer's will likely claim more and more victims in the years ahead. In developed nations, an estimated 2% of the population currently has the disease, and one study projects three times more by 2054. If such predictions are accurate, the potential human suffering and attendant financial burdens will be staggering.

The modern emergence of Alzheimer's may represent a confluence of several separate human health trends: improved public-health standards, development of novel medicines to combat infectious diseases, high-calorie diets rich in saturated fats, and general lifestyle changes. While reversing all these trends is impossible, understanding the factors underlying the emergence of Alzheimer's is now a vital facet in the effort to manage this dementia.

First and foremost, people live much longer than they did a century ago, and Alzheimer's is overwhelmingly an affliction of the aged. From 1900 to the present, the mean life expectancy for a U.S. resident increased from around 50 years to nearly 80 years. While the incidence of Alzheimer's is roughly 1% among 70-year-olds, it is 39% among those 90 to 95 years old.

Although it is cold comfort, the emergence of Alzheimer's is a consequence of sustained and successful efforts toward improving health that are enabling many more people to live long enough to develop dementia. People today have far better protection against once-rampant infectious diseases that ensured many an early death. Improved public sanitation, clean drinking water, effective antibiotics, widespread vaccination, and other measures have all contributed to substantially healthier lives.

Coupled with this huge success against many infectious diseases have been enormous alterations in people's nutrition habits and physical activity. Diets increasingly rich in calories and fats have been accompanied by a declining need for physical labor due to mechanization, the shift toward a service economy, and greater reliance on computerized information. In general, jobs demanding strenuous physical labor have decreased, while sedentary white-collar work has increased. So higher calorie consumption and less calorie-burning activity have created a population that is, on average, more overweight and less physically fit than ever. Obesity is linked to a significantly increased propensity to develop diabetes, heart disease, and hypertension. And recent research suggests that chronic circulatory conditions carry an increased risk for Alzheimer's development, as well.

TREATING ALZHEIMER'S DISEASE

Early-stage Alzheimer's patients can now be treated with drugs that improve cognitive function by allowing neuron chemical signals to last a bit longer in the brain. Although often of great benefit, this treatment is not a cure, and all Alzheimer's patients exhibit an irreversible mental capacity loss that follows a years-long process of neural cell destruction. Postmortem examination of Alzheimer's patient brains has revealed a striking correlation between unique neural tissue abnormalities and this dementia.

The patients accumulate large deposits of a peculiar small protein—amyloid—within and between the neurons and in the walls of blood vessels supplying the brain. Amyloid is formed when a much larger precursor protein is cut into small fragments by enzymes, and it might actually protect the brain by sealing blood vessel leaks that arise through trauma or aging, according to Alex Roher of the Sun Health Research Institute. Although the exact function of amyloid is unknown, the fact that this protein and its larger precursor molecule have been conserved evolutionarily suggests they are important.

In Alzheimer's disease, excessive amyloid accumulates to create the characteristic insoluble plaques and degenerating neurons called tangles in regions of the brain that are essential for higher-order intellectual functions and memory generation. Amyloid proteins are toxic, and plaques are often surrounded by a halo of dead and malfunctioning neural cells. In addition to the spectacular amyloid deposits, it is now apparent that invisible pools of soluble amyloid lurk unseen in Alzheimer's patients' brain tissue, and these molecules may be toxic as well. Dementia may be the final outcome of neurons succumbing slowly to an increasingly toxic environment.

For more than a decade, Alzheimer's research has been driven by the "amyloid hypothesis," which postulates that it is the massive deposits of the protein in brain tissue and blood vessels that causes neurodegeneration. The amyloid hypothesis is appealing because it links the most spectacular brain pathology features to dementia, and it has become almost a dogma among many Alzheimer's researchers.

If we could control amyloid production or remove deposits, some researchers believe, perhaps we could prevent Alzheimer's disease. Recent research with transgenic animals offers hope, though tests on human subjects have encountered difficulties, such as severe brain inflammation among some Alzheimer's patients undergoing amyloid vaccination trials.

The cause(s) underlying the most common form of Alzheimer's may be more subtle than the accumulation of a protein leading to brain-tissue damage. The amyloid plaque deposits might not even be pathological, but rather part of a defense mechanism to make excess amyloid safe by confining it to a sort of "dump." It's only when these dumps reach extreme levels that dementia becomes evident.

Other studies pursue potential links between atherosclerosis, a circulatory system disease, and dementia, because factors known to promote atherosclerosis—such as a diet high in calories, saturated fat, and cholesterol—could also impact dementia development risk. In principle, these specific factors are manageable because the patient can control them through either improved diet or drug intervention to adjust circulatory system cholesterol and fat levels.

ENVISIONING THE FUTURE

All forecasts predict higher rates of Alzheimer's disease—bleak forecasts that might even be too optimistic. Several interacting factors are changing simultaneously. Increased obesity rates are already in evidence in the elderly, as well as among much younger people. Factors known to promote weight gain probably also increase the risk of Alzheimer's. It is unclear if early-in-life obesity increases the risk of developing Alzheimer's, but it could actually lower the mean age at which the disease begins to appear—and thus increase the total incidence rate substantially.

Today's treatments provide only temporary relief of symptoms, and there is little prospect for a cure in the near-term future. But researchers now understand much more about the genetic and biochemical pathways that create amyloid associated with Alzheimer's, and this information may lead to ways to control dementia at its source, such as through enzymes that halt toxic amyloid production or through removing deposits by vaccination or other methods. These approaches are promising, but significant problems regarding possible toxicity and collateral tissue damage must be understood fully and addressed before clinical applications are feasible.

If circulatory system disease and Alzheimer's pathology are interrelated, we could potentially find new therapies to forestall, mitigate, or avoid some cases of dementia. Statin drug treatments that lower circulating cholesterol and lipid levels may decrease chances for maladies such as heart attack, stroke, and Alzheimer's simultaneously.

Lifestyle choices plainly influence many disease-development risks, so following commonsense guidelines for health maintenance—eating a balanced diet with limited cholesterol and saturated fats intake, maintaining an appropriate weight, keeping physically fit and mentally active, and so on—is imperative for several health threats, including Alzheimer's. If enough individuals could be convinced to adopt a healthy lifestyle, society as a whole might benefit as Alzheimer's rates and attendant care costs decreased. Simply delaying the typical age of Alzheimer's onset could provide benefits.

Stem cells (cells that are able to [develop] into any type of tissue) offer a tantalizing possibility to reverse aging. Renewing damaged neural tissue would seem to be of obvious benefit, but brain cells must be replaced with precise biochemical duplicates and reconnected accurately so that the original neural network is reconstituted. Rejuvenation efforts that alter cell biochemistry or perturb interneuronal connections might well lead to a wholesale memory loss as devastating as Alzheimer's itself.

In one regard, the future projected by visionaries has arrived. Over the last century, we have doubled the average human life span. But having managed that, we now face new issues, such as the emergence of Alzheimer's disease, that confound our success. The mechanics of Alzheimer's pathology are now being elucidated down to minute genetic and biochemical details, so we can anticipate that this knowledge will lead to future control measures.

Neurologist Kurt Heininger has posed the question, "Is Alzheimer's disease an age-related or aging-related disorder?" The issue becomes critical as we endeavor to further extend the average human life span, because it appears that virtually everyone will get Alzheimer's if we only live long enough. If Alzheimer's is simply the result of age-related accidents, such as insidious vascular system malfunction, we will probably be able to eliminate its root causes and thereby eliminate the specter of this senile dementia.

On the other hand, perhaps Alzheimer's represents one facet of aging-related events that are a part of the unique logic of evolution. Theorists such as Theodore Goldsmith, author of *The Evolution of Aging*, have proposed that genetically programmed death, although dooming the individual, benefits the species. Old generations must give way to the new for evolution to occur, and death may be the ultimate mechanism to guarantee that species undergo continuous change.

If Alzheimer's disease is one component of an evolutionarily conserved programmed aging process, then reaching the visionary's dream of an immensely long human life span may force us to keep defusing some difficult biochemical time bombs one after another.

Neuron Killers[*]

Misfolded, Clumping Proteins Evade Conviction, But They Remain Prime Suspects in Neurodegenerative Diseases

By Tina Hesman Saey
Science News, August 16, 2008

As open-and-shut cases go, Alzheimer's disease should top the list. The victim is clear. Suspects are in custody. Wherever neurons die due to Alzheimer's disease, a protein known as amyloid-beta is always found at the scene of the crime, hanging around in large, tough gangs called plaques. Parkinson's and Huntington's diseases; amyotrophic lateral sclerosis (which goes by its initials ALS or the alias Lou Gehrig's disease); and prion diseases, such as scrapie in sheep, mad cow disease in cattle and Creutzfeldt-Jakob disease in humans, all have similar stories.

Scientific investigators have pieced together this much: A seemingly mild-mannered brain protein falls in with a bad crowd, the corrupted protein and its cronies gang up and mob violence results in the death of a brain cell. It's a scene repeated over and over again in different neighborhoods of the brain, by different proteins, but all with the same result—the death of neurons and rise of disease.

But no one has convicted these suspected neuron killers. So far, cases mostly rely on circumstantial evidence, with large holes in the web of proof. There's no smoking gun, no motive and no eyewitness to corroborate what scientists suspect. And there's no cure for the diseases that slowly break down brains and spinal cords, robbing victims of memories or mobility.

No one has observed all the steps of a neuron's demise, so no one is sure exactly what the murder weapon is or who dealt the final blow. But scientists acting like shamuses on the scent of a killer have picked up tantalizing clues about how neurons meet their end, and protein aggregation is almost certainly involved.

"It seems unlikely that coincidence is at work here," says Bradley Hyman, a neurologist at Harvard Medical School and Massachusetts General Hospital-East in Charlestown, Mass. Recent research from Hyman and colleagues shows that plaques develop more rapidly in the brains of mice prone to Alzheimer's disease than had been thought. The discovery, published February 7 in *Nature*, suggests that there may be many years between the appearance of plaques and the onset of disease, providing a window of time for doctors to take action and stop the death of neurons.

Other researchers have recently reported progress on developing molecules that may help protect the brain against proteins-gone-bad. And other new research shows that the perpetrator in some cases of neurodegenerative disease may not be one of the usual suspects.

The key to stopping the killing of neurons is figuring out what causes otherwise innocuous proteins to show their Mr. Hyde side, and discovering why the proteins flock together once they've turned. The method by which "bad" proteins bump off neurons is also a matter of dispute. Scientists are drawing ever closer to solutions for these mysteries, and what they discover may one day help head off these diseases or even repair some damage after rogue proteins have vandalized the brain or spinal cord.

CAUSE OR EFFECT

Not everyone believes that protein aggregation is such a bad thing for neurons. Take those big plaques of amyloid-beta, or A-beta, found near dead and dying brain cells in Alzheimer's disease patients.

"Some people say it's a tombstone, others say it's not the cause," says Gang Yu, a neuroscientist and biochemist at the University of Texas Southwestern Medical Center at Dallas.

Big clusters of protein may be a cell's way of coping with otherwise harmful proteins, suggests Lila Gierasch, a biophysical chemist at the University of Massachusetts Amherst. Plaques are "like garbage dumps for insoluble proteins," she says. Indeed A-beta plaques contain remnants of other proteins, perhaps dumped in the plaque to avoid cluttering up a cell and gumming up its inner workings.

Brain images of healthy people reveal that A-beta plaques are common, even in people who don't have dementia. And mice that make a lot of A-beta have memory problems, but their neurons don't die, says Li-Huei Tsai, a neuroscientist at MIT. "The role of A-beta is still very, very controversial," she says. Some people think elevated levels of the protein may interfere with neuron communication. Others think that small aggregates, rather than large clumps, are toxic to cells.

Part of the difficulty in deciphering A-beta's role in Alzheimer's disease is that no one is sure what the protein's day job is. That's true of alpha-synuclein, a protein that forms clumps called Lewy bodies inside brain cells of people with Parkinson's disease, and of huntingtin, a protein which has been shown to be the causative agent of Huntington's disease. Alpha-synuclein, A-beta and the prion protein PrP probably aren't unemployed, but scientists have not yet established their roles.

On the surface, these proteins, as well as two proteins (TDP-43 and superoxide dismutase or SOD1) involved in ALS, have nothing in common, says Mark Goldberg, director of the Hope Center for Neurological Disorders at Washington University in St. Louis. The sequences of amino acids that compose the proteins aren't the same, nor are the normal shapes of the proteins. The neuron-killing proteins probably function differently too. But all of them go bad in a similar way, twisting from loose, flexible molecules into rigid, sticky formations known as beta-pleated sheets.

Every protein in the body probably has the ability to form beta-pleated sheets given the right (or wrong) circumstances, says Erich Wanker, a molecular biologist and biochemist at the Max Delbrück Center for Molecular Medicine in Berlin. Something about these proteins and others that cause amyloidosis—fatal diseases in which abnormally folded proteins build up in organs—makes the proteins more prone to assuming the deadly conformation. Genetic mutations can tip the balance, but that doesn't explain why people who don't have mutations sometimes end up with the aggregates.

ON THE STRAIGHT AND NARROW

Although the precipitating event that leads good proteins down the beta-pleated path isn't known, Wanker and his colleagues may have developed a way to stop the process, at least in the test tube. In a report published in the June *Nature Structural & Molecular Biology*, Wanker and his collaborators showed that a small molecule called (—)-epigallocatechin gallate (mercifully shortened to EGCG) can keep A-beta and alpha-synuclein from forming beta sheets. The group had previously shown that the compound could prevent huntingtin from aggregating.

EGCG latches on to the backbones of the amino acid chains that compose the proteins. With EGCG riding piggyback, the proteins form small clumps. But apparently the proteins never switch to the beta-sheet formation, so the little clumps aren't toxic to cells in the test tube.

Wanker doesn't know whether EGCG, found in green tea, would be an effective therapy for neurodegenerative diseases. The researchers have yet to demonstrate that the compound can dissolve existing aggregates. Also, the experiments used equal parts of the molecule to protein in order to stop the proteins from forming the toxic beta sheets, which may mean that therapies would require massive amounts of the compound to work effectively. It's also not known how well EGCG gets across the blood-brain barrier. If the molecule doesn't enter the brain easily, doses of EGCG needed to prevent disease might be too high to be practical.

Cells may already possess molecules that work in the same way EGCG does, Wanker says. Proteins called chaperones also help keep other proteins loose and ready for action. Some evidence suggests that defects in chaperones may be the blow that sets off brain-wasting diseases. "This mechanism may be more common than we think," Wanker says.

Other proteins may act as guardian angels to keep would-be neuron killers on the straight and narrow too. One such guardian may be a protein known as Pin1, which could keep another potential killer that stalks the brains of Alzheimer's disease patients from turning deadly.

While spotlights have been trained on A-beta as the most likely killer of neurons in Alzheimer's disease, Kun Ping Lu of Beth Israel Deaconess Medical Center in Boston thinks scientists may be ignoring a more deadly culprit, a protein called tau.

Tau is normally a hard-working protein that helps create the internal skeleton of the cell by binding to the cell's frame-supporting microtubules. If not for tau, the long fibers called axons that connect neurons across the brain would break down, severing communication as surely as cutting a fiber-optic cable to a building would. Dendrites, the neuron's branchlike projections that receive signals from other neurons, would also disintegrate without tau pinning microtubules in place.

People who have mutations in the gene that encodes tau develop a disease called frontotemporal dementia. The brains of people with this dementia look much like brains of people with Alzheimer's disease with one critical difference: Frontotemporal dementia patients don't have plaques in their brains. But they do have tangles of tau in brain cells, and their neurons are as dead as [those in] a person with Alzheimer's disease.

That leads Lu to believe that tau may be more directly involved in killing neurons than A-beta. In other words, A-beta may order the hit, but tau pulls the trigger. "If, on top of tangles, you add plaques or increase A-beta, now you have massive neurodegeneration," Lu says.

Lu lays out the scenario for brain-cell murder this way: A-beta builds up outside neurons, leading to inflammation in the brain. Inflammation prods enzymes called kinases to tack extra phosphates on to tau inside the cells. This causes tau to walk off the job and hang out in hard tangles with other tau molecules that have more phosphates hanging off them than groupies on a rock star. Hyperphosphorylated tau forms such tight bonds with its cronies, not even boiling it in detergent can untangle it, Lu says. After that, it's all over for the neuron as its axons and dendrites collapse.

Normally, tau's protector, Pin1, keeps it from falling in with hardened tangles. Pin1 actually does double duty, watching over tau and APP, the protein precursor to A-beta. Mutations in the gene for Pin1 have now been linked to late-onset Alzheimer's disease, but not to early onset forms.

Lu and his colleagues have found a variation in the Pin1 promoter, a stretch of DNA that controls activity of the gene, associated with a five-year later onset of Alzheimer's disease. The researchers don't yet know if the variation increases Pin1 production. They do know that aging causes Pin1 production to fall.

"As people get older and older, Pin1 levels drop, drop, drop," Lu says.

Boosting Pin1 levels may help untangle tau in people at risk of Alzheimer's disease, slowing the disease's progression or preventing it altogether. Reporting in the May *Journal of Clinical Investigation*, Lu and colleagues showed that making more of the protein could help protect against tangle formation in mice. But the new research also shows that too much Pin1 can be a bad thing. When researchers increased Pin1 levels in mice carrying the P301L alteration in tau—found in people with frontotemporal dementia—more brain cells died than did in mice that carry the tau mutation but make normal levels of Pin1.

THE POISONING BLAME GAME

Tau is not the only protein that may be getting away with neuron murder while a more high-profile suspect takes the rap. The antioxidant protein superoxide dismutase had been fingered as the killer of spinal cord neurons in people with ALS. A small subset of those with the disease have mutations in the gene for SOD1 that lead to clumping of the protein and the death of neurons that direct motion.

But recently scientists learned that nearly everyone with ALS has aggregates of a protein called TDP-43 (for TAR DNA binding protein) in their spinal neurons.

"If TDP-43 is the major pathway, then SOD1 was misdirecting us," says Christopher Shaw, a neurologist and neurogeneticist at King's College London. He estimates that about 1 percent of people with ALS have mutations in the gene for TDP-43. Shaw and his colleagues showed in a report published March 21 in *Science* that those mutations lead the protein to stick together more readily. Most cases are sporadic, not inherited, and occur when TDP-43 twists into a shape that favors aggregation. Scientists don't yet know what sets off the conversion, but Shaw says the tail of the molecule certainly plays a role.

The tail end of TDP-43, what scientists refer to as the c-terminus, "aggregates fantastically quickly," he says. "It's an extremely sticky little beast."

That stickiness is characteristic of all proteins that form neuron-killing beta sheets and may account for the speed at which plaques and other aggregates form. Although scientists have evidence that proteins become toxic after twisting into beta sheets and aggregating, just how clumps of protein poison neurons isn't clear.

For instance, even though SOD1 protein is made everywhere in the body, and mutations that lead to overproduction cause aggregation of the protein in many tissues, only spinal cord neurons degenerate to give rise to ALS. Similarly, neurons that produce dopamine, a chemical key to neural communication, are the victims of alpha-synuclein clumps in people with Parkinson's disease. And Alzheimer's plaques tend to congregate in parts of the brain that are active when people are daydreaming or thinking about nothing in particular.

The life cycle of a neuron might explain its susceptibility to damage, Shaw says. Most neurons last a lifetime. The cells don't divide after they are born and take their place in the brain. Some new neurons do develop in parts of the brain, but most of the 10 billion to 100 billion neurons are present before birth and last until death. The cells never get a day off and they have no backup or replacement.

Their long lives may lead neurons to produce proteins differently than other cells. "Maybe brain cells have a just-in-time policy," Shaw says. "You don't make a lot of protein and stack it up, so therefore you don't have the same rigorous protein turnover mechanisms." In other cells in the body, quality control would quickly recognize a misfolded protein and get rid of it before it could cause mischief. The lack of supervision in neurons could make them more vulnerable to rogue proteins.

On the other hand, neurons may process proteins correctly, but age may catch up with the neurons, making them weary of the constant effort against aggregation.

"There's an ongoing battle for many years, and ultimately the neuron gives up," speculates Yu from UT Southwestern. But scientists don't know what causes neurons to throw in the towel. The final straw could be the loss of chaperone proteins, which oversee protein-folding, or a strike by the cellular machinery that transports or breaks down proteins, causing crowding in the cell that foments aggregation.

"Theories abound," Yu says, "but none have been definitively proven."

Keeping Your Brain Fit[*]

By Christine Larson
U.S. News & World Report, February 11, 2008

Marian Conte's brain weighs 1,100 grams, according to Nintendo. "That's up from 800 grams when I started playing," jokes Conte, 52, a real-estate agent from Hamilton, N.J., who recently added the video game Big Brain Academy to her fitness regimen. The better she scores on brainteasers, the larger her fictional brain. Since Conte's mother died of complications from Alzheimer's disease in 2003, she's trying to guard herself any way she can, embracing crossword puzzles, fruits and vegetables, and a new genre of high-tech workouts that aim to slow cognitive loss. This particular game makes no such claim. But regular play certainly can't hurt, Conte figures: "I want to do any little thing I can to protect my brain."

THE INCREDIBLE SHRINKING BRAIN

If her Nintendo score isn't solid evidence, science increasingly suggests Conte's efforts may pay off. Just within the past few months, several groups of researchers have added support for the growing consensus that plenty can be done to slow the age-related declines in memory, mental speed, and decision making that affect most people. In November, a team from the Mayo Clinic and the University of Southern California announced that one computer-based mental training program appeared to improve older people's cognitive performance by as much as 10 years. That same month, a Harvard researcher found that long-term use of beta carotene supplements delayed cognitive decline by up to a year and a half.

And a new book out last month puts forth evidence that "exercise is the single best thing you can do for your brain," says author John Ratey, a clinical associate professor of psychiatry at Harvard Medical School. The book is *Spark: The Revolutionary New Science of Exercise and the Brain.*

"Some of the myths about the brain—that it was not changeable, that there was nothing you could do about cognitive decline—have really been dispelled in the past 10 years," says Lynda Anderson, director of the Healthy Aging Program at the federal Centers for Disease Control and Prevention, whose bold goal is "to maintain or improve the cognitive performance of all adults." The potential payoff is enormous. Alzheimer's now afflicts 4.5 million people in the United States—double the number in 1980—and is expected to reach 16 million by 2050. "Statistics show if we could delay the onset of Alzheimer's by five years, the number of people with the disease would be cut in half," says Yaakov Stern, a cognitive neuroscientist at Columbia University.

What are you up against? The inevitable physical changes start in early adulthood but become especially marked after about age 60 or so. Gradually, the brain shrinks, losing around 0.5 percent to 1 percent of its volume each year after that age threshold; brains with Alzheimer's shrink about twice as fast. The effects are greatest in the prefrontal cortex, the seat of executive function (which includes working memory—responsible for remembering a telephone number while you're dialing, say—and planning, focus, and behavior choices), and sometimes in the hippocampus, involved in memory. Brain cells' dendrites and axons—the slender filaments that transmit electrical impulses—shrink. The brain's white matter, which contains nerve fibers that transmit signals from one brain region to another, starts to degrade around age 50. Result: It gets harder and harder to remember what you wanted to buy at the grocery store, to process and respond to information, and to reason your way through a problem. In your 70s and 80s, executive function starts to fail.

Not every mental skill suffers equally. Vocabulary, for instance, tends to remain, as do skills practiced for a long time, like playing the piano or using a spreadsheet. You might even improve at some things: In tests of experienced crossword puzzlers of all ages, the best were in their 60s and 70s.

Potential. The more scientists learn about the brain's decay, the more curious they've become about how well people function anyway. Even among people 85 and older, only 18.2 percent live in nursing homes. "In the past, much of the research has focused on disease and decline," says Gene Cohen, director of the Center on Aging, Health and Humanities at George Washington University. "Now we're looking at the concept of potential and how older people often continue to thrive and grow even in the face of the most serious illness." Recent studies of both animal and human subjects have found that several factors go hand in hand with better mental performance, including education, professional success, and intellectual, social, and physical activities. A 2003 study reported in the *New England Journal of Medicine*, for example, found that people over 75 who danced, read, or played board games or musical instruments also had a lower rate of dementia.

Much of the work has focused on finding ways to bulletproof people against Alzheimer's. In mice, an Alzheimer's vaccine seemed to work, but it proved toxic in humans and trials were suspended (although research on vaccines continues). Beta carotene supplements may delay cognitive decline if taken for many years—but only by a year and a half. Education seems to lower your odds of Alzheimer's—

but even some Nobel laureates develop it. Cholesterol-lowering drugs seemed to offer some promise in fending off Alzheimer's, but a 12-year-long study published in January showed they had no effect. For now, experts think the best approach is to take the sorts of steps that Conte is taking to delay normal cognitive decline.

Stretch the plastic. For decades, scientists assumed that humans were born with all the brain cells they'd ever have. Then, in the 1970s, researchers showed that new brain cells and neural pathways form through the end of life. "This was the beginning of the brain plasticity movement," says Cohen, "the understanding that when we challenge our brains, the brain cells sprout new dendrites, which results in increased synapses, or contact points." More recent research has shown that there isn't an age limit: Training older adults in certain memory tasks, like remembering faces and names, seems to boost those specific abilities—though it won't remind you to bring your shopping list to the store. And the newest evidence suggests that intensive practice in reasoning skills or in distinguishing sounds appears to lead to more generalized improvements in brain function.

In 2006, for example, a controlled clinical study of more than 2,000 older people by researchers at Pennsylvania State University, Indiana University, Johns Hopkins University, and elsewhere found that those who received 10 60-to-75-minute training sessions in reasoning—specifically, in recognizing word, number, and letter patterns and filling in the next item in a series—reported less difficulty with such activities of daily living as understanding instructions on a medication label. The effects still were apparent five years later. This past November, scientists from the University of Southern California and the Mayo Clinic announced that study subjects who spent an hour a day for eight to 10 weeks using a program that asked them to recognize subtle differences in sounds performed better than the control group on memory and speed tests, too. Designers of the Brain Fitness Program (made by Posit Science, which funded the study) claim that such ear training causes the brain to convey information more precisely from one region to another—which, in turn, improves other types of thinking.

"The amount of memory improvement was equivalent to going back 10 years in your ability," says Elizabeth Zelinski, professor of gerontology and psychology at USC and a principal investigator on the study, which has not yet been published.

Experts caution that most brain-training products haven't been tested and that what data do exist are still shaky. If improvement of daily living tasks is the goal, "we don't yet have the data to suggest they accomplish that," says Arthur Kramer, a neuroscientist at the University of Illinois. "Yes, we have data that says you can get better at certain things with practice. But does it translate to the real world? We don't know yet." Still, many doctors who work with older people feel they don't have time to wait for the research, and nursing homes and senior centers across the country are adding "brain gyms" and other programs to help older people stay mentally active.

"I've learned more about China than you can imagine," says Hortense Gutmann, 100, who started using E-mail just over a year ago through a new computer-education program for residents of Sarah Neuman Center for Healthcare and Rehabilitation, a nursing home in Mamaroneck, N.Y. She now keeps in touch with relatives there, as well as in Minnesota and Israel, and takes great pleasure in having mastered a new skill.

Consumers aren't waiting for more research, either. The market for products like Brain Fitness Program, Nintendo's Brain Age, and MindFit soared to an estimated $80 million in 2007, up from just $2 million to $4 million in 2005, according to SharpBrains.com, a San Francisco-based group that follows the industry. Meanwhile, the Alzheimer's Association recommends any activity that will keep you curious and learning: reading and writing, attending lectures, taking classes, even gardening.

Sound body, sound mind. Still, the best workout for your brain may be the old-fashioned kind.

As far back as 1999, researchers at the University of Illinois found that older people who started exercising showed faster reaction times and better ability to focus after just six months than did a control group. Now, it's becoming clearer why. In a second study reported in 2006, the same team found that the aerobic exercisers actually increased their brain size by about 3 percent. Last year, researchers at Columbia University found that when people exercised regularly for three months, blood flow increased to a part of the hippocampus, which is important for memory. In studies of mice who exercised on treadmills, increased blood flow to the same part of the brain corresponded with an increase in the production of new brain cells.

The power of exercise seems far more impressive than that of brain-training software, says Sandra Aamodt, editor in chief of *Nature Neuroscience*, a scientific journal on brain research, and coauthor of the forthcoming book *Welcome to Your Brain*. A recent meta-analysis of numerous exercise studies found that, on average, faithful aerobic exercise might boost someone's cognitive performance from average—say, from 10th place out of 20 people tested—to notably above average—say, to No. 5. But cognitive training would boost the same person to eighth out of 20.

Why is exercise so good for the brain? Maybe for the same reason it's so good for the heart: its beneficial effect on blood vessels. "It may be that a pretty significant amount of deterioration in brain function relates to disruptions of the cardiovascular system by microstrokes," in the tiny vessels in the brain, says Aamodt. Exercise may help prevent them. It also stimulates the production of proteins called growth factors, which promote the formation and growth of brain cells and synapses.

Certain nutrients, too, are thought to be protective. The antioxidants in fruits and vegetables have been linked to improved cognitive function; berries, for instance, seem especially beneficial in keeping brains spry. "Old neurons, like a lot of old married couples, don't talk to each other anymore," says James Joseph, director of the neuroscience lab at the USDA Human Nutrition Research Center on Aging at Tufts University. "We have found that the berry fruits improve neuronal communication." In November, Harvard researchers announced that men who took a beta carotene supplement for 18 years had slightly better cognitive function than those who didn't—their memory scores matched those of people about one year younger. However, men who took supplements for only one year showed no improvement, and several other studies have found no link between antioxidants and mental performance. The Alzheimer's Association recommends a diet high in dark-colored veggies, like kale, spinach, beets, and eggplant; colorful fruits like berries, raisins, prunes, oranges, and red grapes; plus fish like salmon or trout high in heart-healthful omega-3 fatty acids.

Making connections. It has been more than two decades since Bill Harves, 90, quit singing in his church choir. Four years ago, he joined the professionally led chorale that rehearses once a week at his Bailey's Crossroads, Va., continuing care retirement community. The chorale gives several concerts a year, including one at Washington, D.C.'s Kennedy Center. He's gained in breathing technique, enunciation, and music reading skills. "There's no doubt I've improved as a singer," he says.

Besides having fun, Harves, who also serves as chairman of his community's computer club and is active on a residents' committee, is very likely protecting his cognitive function. In a study of more than 2,800 people ages 65 or older, Harvard researchers found that those with at least five social ties—church groups, social groups, regular visits, or phone calls with family and friends—were less likely to suffer cognitive decline than those with no social ties.

"The working hypothesis is that it has something to do with stress management," says Marilyn Albert, a neuroscientist at Johns Hopkins and codirector of the Alzheimer's research center there. In animal studies, a prolonged elevation in stress hormones damages the hippocampus. Social engagement appears to boost people's sense of control, which affects their stress level. Creative arts seem to be a highly promising way to increase social engagement. George Washington University's Cohen has found that elderly people who joined choirs also stepped up their other activities during a 12-month period, while a nonsinging control group dropped out of some activities. The singers also reported fewer health problems, while the control group reported an increase.

All the new research has senior programs rethinking their offerings. In Chicago, for example, Mather LifeWays, a not-for-profit that promotes healthful aging, has opened three neighborhood cafes that serve coffee and sandwiches to people of all ages and offer fitness classes, computer courses, lifelong-learning opportunities, and volunteer activities for older adults. "I've met lots of friends here," says Jill Wonsil, 66, who drops in at the cafe near her home several times a week to socialize, check E-mail, and take exercise and other classes. If living life to the fullest is the best way to stay sharp, it's not such a tough prescription to swallow.

6

The "Hard Problem": Efforts to Understand
Human Consciousness

Editor's Introduction

Even as scientists continue to make great strides mapping the brain and learning the functions of its various structures, one perplexing question remains: How do we explain consciousness? The problem is made all the more complicated by the fact that many scientists have differing views of what the term even means. Broadly defined, consciousness is the "I" inside of our minds, the self-awareness that leads us to believe we're in control of our actions and experiencing the world in our own unique ways. The Greek philosopher Plato believed the mind and body to be two separate entities, a "dualistic" view that is roughly analogous to the concept of the soul, a central tenet of many of the world's religions. Few modern scientists subscribe to such "ghost in the machine" beliefs, and instead, they search for consciousness in the chemistry of the brain itself.

Attempting to frame the question of how to explain consciousness, the philosopher David Chalmers has posited that modern science must solve two problems, one easy, one hard. The so-called Easy Problem—which, it turns out, isn't so easy—is to differentiate between conscious and unconscious mental functions and determine how they are processed in the brain. The Hard Problem, on the other hand, is to explain how our experiences—everything we see, touch, taste, hear, feel, think, etc.—add up to create our sense of self. "The problem is hard because no one knows what a solution might look like or even whether it is a genuine scientific problem in the first place," the scientist and author Steven Pinker wrote for *Time* magazine. It's for this reason that one subset of scientists believes the essence of human consciousness will forever remain a mystery.

The articles in this chapter provide an overview of how scientists and philosophers are tackling consciousness, perhaps the most vexing question mankind has ever faced. In the first entry, "What Makes Up My Mind?" Joel Achenbach introduces concepts such as dualism and Chalmers' Easy and Hard Problems. He also makes reference to the writing of Daniel Dennett, an author and philosopher who believes our brains are similar to automobiles that drive themselves. In Dennett's version of consciousness, there is no "president in the Oval Office of the brain," but rather a number of competing states of awareness. In "A User's Guide to the Brain," Pinker expounds on many of the aforementioned theories, considering their moral and philosophical implications. While he casts his lot with those scientists who believe that our brains are simply not equipped to comprehend something as complex as consciousness, he holds out hope that a Darwin- or Einstein-like figure might one day present a "flabbergasting new idea that suddenly makes it all clear to us."

In her piece "Brains Wide Shut?" Patricia Churchland criticizes many of the books that have been written on consciousness, arguing that they have contributed little to the debate. She focuses on Dennett's theory that, at any given moment, consciousness represents one set of sensory signals temporarily taking control of the brain, challenging his assertion with scientific findings. Churchland concludes the article by opining that scientists won't truly understand consciousness until they've completed the long "empirical slog" of learning how each part of the brain—down to the molecular level—functions. In "Is There Room for the Soul?" Jay Tolson catalogs many of today's leading theories on consciousness. While he, too, cites the work of Dennett, he also discusses those thinkers who believe the brain is more than just an autonomous "Darwinian survival machine." In the final selection Bruce Bower considers whether consciousness might reside in the brain stem, a theory that has resulted from studies of children born without portions of their cortex. In many instances these children show basic signs of consciousness, even though they live in vegetative states.

What Makes Up My Mind?*

By Joel Achenbach
Washington Post, September 23, 2007

If I were to be eaten by a shark, I'm pretty sure the worst part would be not the pain or the mutilation or the actual dying and so forth, but rather the thought balloon over my head with the words, "I'm being eaten by a [expletive] shark!"

Whereas a fish doesn't have this problem. A fish has no thought balloon, or just a teensy little one, with a monosyllabic fish-word like "Urp!" A fish probably suffers, but it doesn't have the additional suffering that comes from knowing that it's suffering, and from regretting that it went swimming instead of watching the golf tournament, and from hearing, as we all do whenever we're devoured by sharks, the theme music from "Jaws." You know: that tuba.

All of which is a deft way of introducing our subject today: The Mystery of Consciousness. It's one of the biggest unknowns, right up there with the origin of life. But it's under a multi-pronged assault by scientists, who vow to crack the code of the mind in the same way that they are deciphering the human genome. It's all very exciting, with the one catch that no one can really agree on what the mind is.

"With consciousness, there is no agreement on anything," says Giulio Tononi, a professor of psychiatry at the University of Wisconsin at Madison, "except it's very difficult."

Jim Olds, who directs George Mason University's Krasnow Institute, a think tank devoted to the study of the mind, says of his field, "We're waiting for our Einstein."

The human brain is a hunk of meat that weighs about three pounds. It contains about 30 billion cells, called neurons. The networking of these cells involves 100 trillion meeting points, or synapses. This is the most complex object in the known universe (though if we explore the stars we may eventually find organisms with brains that make ours seem as impressive as Twinkies).

Human brains can do things that no computer can match. Sure, a computer can beat a human at chess, but only with brute-force calculation of every conceivable move. The most sophisticated robots still lack the basic smarts of a 2-year-old, who can perceive the world in three dimensions and go searching for a kitty cat while somehow avoiding the jutting edge of the coffee table. Negotiating the world requires massive bandwidth.

"The engineering problems that we humans solve as we see and walk and plan and make it through the day are far more challenging than landing on the moon or sequencing the human genome," psychologist Steven Pinker writes in his book "How the Mind Works."

Beyond the basics of perception and motor skills, the human brain has a premium feature: consciousness. You could also call it sentience, or self-awareness, or just the thing that makes it such a drag to be devoured by a mindless oceanic carnivore. This is what keeps us from being zombies. We perceive ourselves as actors on the stage of life. We sense that there's an "I" somewhere inside our skull.

"Consciousness is a big thing," Tononi says. "It is the single biggest thing of all. It is the only thing we really care about in the end."

But we don't understand it. We don't know how, in the words of philosopher Colin McGinn, "the water of the physical brain is turned into the wine of consciousness."

WILL WE EVER KNOW?

Earlier this year, Jim Olds gathered a bunch of big thinkers at George Mason University for a two-day conference on the mind. He and his allies want the federal government to invest $4 billion in an initiative that would be called the "Decade of the Mind." This would be a follow-up to a 1990s program called the "Decade of the Brain," which brought increased attention to neuroscience. The new initiative would be an attempt to take science into a realm previously explored only by philosophers, theologians and mountaintop yogis.

"Brain science is an exhaustive collection of facts without a theory," Olds says. "This is for the nation as a whole to invest in one of the fundamental intellectual questions of what it is to be a human being."

In a letter published a few weeks ago in the journal *Science*, 10 scientists said that a Decade of the Mind would help us understand mental disorders that affect 50 million Americans and cost more than $400 billion a year. It might also aid in the development of intelligent machines and new computing techniques. A breakthrough in mind research, the scientists wrote, could have "broad and dramatic impacts on the economy, national security, and our social well-being."

There's reason to be optimistic. Look at what has happened in recent years with the development of brain scans, such as MRIs, that let us observe the brain at work in real time. As the technology improves, the brain becomes more transparent, less of a black box.

That said, the mind isn't something that pops up on a computer screen. People have been poking around the brain in search of the mind for many centuries, and no one is even sure what neurological structures are the most critical to generating consciousness. Descartes, who gave us the most famous line in the annals of philosophy ("Cogito, ergo sum"—I think, therefore I am), believed the center of consciousness to be the pea-size structure known as the pineal gland. Nice stab, but it turns out that the pineal gland does not seem to have much to do with creating the "I" in our head.

Other brain structures are important, such as something called Brodmann area 46, and the anterior cingulate sulcus, and the thalamus, and of course the knurled, dipsy-doodle structure called the cerebral cortex. We can also be confident that consciousness does not depend on the cerebellum, which is 50 billion neurons worth of brain matter that you could surgically remove without "losing your mind." As Tononi puts it, you could toss the cerebellum in the garbage and " you would still be there."

The classic idea of "dualism" solves the location problem by defining it away: The mind is perceived as separate from the body, something that can't be reduced to machinery. It's unreachable by the tools of the laboratory. Dualism flatters us, for it suggests that our minds, our selves, are not merely the result of rambunctious chemistry, and we are thus free to talk about souls and spirits and essences that are unfettered by the physical body.

Dualism is pretty much dead to serious researchers, though an echo of it can be found among philosophers who are sometimes called the Mysterians. The philosopher David Chalmers has famously made a distinction between the Easy Problems, which involve the ways that the brain creates specific elements of consciousness (vision, language, memory, attention, emotion, etc.), and the Hard Problem, which is the mystery of how all the elements come together in that powerful sense of self ("I am Spartacus").

But here's the most radical idea of all: The reason why the mind is hard to define is not because it has some mysterious, ethereal, spooky qualities but because it doesn't really exist. We just imagine it. You might say it's all in our heads.

When you see a Toyota cruising down the street, you know that you're looking at a complex machine with many parts. You also know that there's a person inside, some intelligent being who's directing the Toyota's movements. The human brain is another complex machine with many parts—but it doesn't seem to have a driver most of the time.

The brain operates day and night and performs myriad functions of which we have no direct awareness. Even our "conscious" brain is actually many different operating systems. It's as though the Toyota is being driven by hundreds of tiny elves, with no single elf in charge.

This is the view espoused by the philosopher Daniel Dennett, author of "Consciousness Explained," who argues that the notion of a central executive in the brain is an illusion. "It's a mistake to look for the president in the Oval Office of the brain," he declares.

It's bad enough that astronomers tell us that the Earth isn't at the center of the cosmos; it's worse that biologists tell us we're all descended from pond scum. Now we have philosophers saying that the self is illusory. You are not really there.

The mind might be what Pinker calls the "ultimate tease." He has written that "the most undeniable thing there is, our own awareness, would be forever beyond our conceptual grasp."

The mind, in this view, isn't a single, specific thing. It's more like a process, or an "emergent" phenomenon. This means that the many disparate components are not themselves conscious, but when they get together, the consciousness precipitates into being. Grabbing hold of the mind, however, would be like trying to seize a puffy white cumulus cloud.

Cracking the code of the mind may be ultimately impossible. My guess is that a century from now, consciousness will still make the list of Biggest Mysteries and scientists and philosophers will still be arguing about the what, where and how of it all.

But we should still take a whack at it. Ten years and $4 billion: That's a reasonable cost. The evolution of the human mind is arguably the most important biological event in the history of our planet since the origin of life itself.

We should try to understand how the brain makes the mind. And then we can make up our minds about what to do with ourselves.

The Mystery of Consciousness[*]

By Steven Pinker
Time, January 19, 2007

The young women had survived the car crash, after a fashion. In the five months since parts of her brain had been crushed, she could open her eyes but didn't respond to sights, sounds or jabs. In the jargon of neurology, she was judged to be in a persistent vegetative state. In crueler everyday language, she was a vegetable.

So picture the astonishment of British and Belgian scientists as they scanned her brain using a kind of MRI that detects blood flow to active parts of the brain. When they recited sentences, the parts involved in language lit up. When they asked her to imagine visiting the rooms of her house, the parts involved in navigating space and recognizing places ramped up. And when they asked her to imagine playing tennis, the regions that trigger motion joined in. Indeed, her scans were barely different from those of healthy volunteers. The woman, it appears, had glimmerings of consciousness.

Try to comprehend what it is like to be that woman. Do you appreciate the words and caresses of your distraught family while racked with frustration at your inability to reassure them that they are getting through? Or do you drift in a haze, springing to life with a concrete thought when a voice prods you, only to slip back into blankness? If we could experience this existence, would we prefer it to death? And if these questions have answers, would they change our policies toward unresponsive patients—making the Terri Schiavo case look like child's play?

The report of this unusual case last September was just the latest shock from a bracing new field, the science of consciousness. Questions once confined to theological speculations and late-night dorm-room bull sessions are now at the forefront of cognitive neuroscience. With some problems, a modicum of consensus has taken shape. With others, the puzzlement is so deep that they may never be resolved. Some of our deepest convictions about what it means to be human have been shaken.

It shouldn't be surprising that research on consciousness is alternately exhilarating and disturbing. No other topic is like it. As René Descartes noted, our own consciousness is the most indubitable thing there is. The major religions locate it in a soul that survives the body's death to receive its just deserts or to meld into a global mind. For each of us, consciousness is life itself, the reason Woody Allen said, "I don't want to achieve immortality through my work. I want to achieve it by not dying." And the conviction that other people can suffer and flourish as each of us does is the essence of empathy and the foundation of morality.

To make scientific headway in a topic as tangled as consciousness, it helps to clear away some red herrings. Consciousness surely does not depend on language. Babies, many animals and patients robbed of speech by brain damage are not insensate robots; they have reactions like ours that indicate that someone's home. Nor can consciousness be equated with self-awareness. At times we have all lost ourselves in music, exercise or sensual pleasure, but that is different from being knocked out cold.

THE "EASY" AND "HARD" PROBLEMS

What remains is not one problem about consciousness but two, which the philosopher David Chalmers has dubbed the Easy Problem and the Hard Problem. Calling the first one easy is an in-joke: it is easy in the sense that curing cancer or sending someone to Mars is easy. That is, scientists more or less know what to look for, and with enough brainpower and funding, they would probably crack it in this century.

What exactly is the Easy Problem? It's the one that Freud made famous, the difference between conscious and unconscious thoughts. Some kinds of information in the brain—such as the surfaces in front of you, your daydreams, your plans for the day, your pleasures and peeves—are conscious. You can ponder them, discuss them and let them guide your behavior. Other kinds, like the control of your heart rate, the rules that order the words as you speak and the sequence of muscle contractions that allow you to hold a pencil, are unconscious. They must be in the brain somewhere because you couldn't walk and talk and see without them, but they are sealed off from your planning and reasoning circuits, and you can't say a thing about them.

The Easy Problem, then, is to distinguish conscious from unconscious mental computation, identify its correlates in the brain and explain why it evolved.

The Hard Problem, on the other hand, is why it feels like something to have a conscious process going on in one's head—why there is first-person, subjective experience. Not only does a green thing look different from a red thing, remind us of other green things and inspire us to say, "That's green" (the Easy Problem), but it also actually looks green: it produces an experience of sheer greenness that isn't reducible to anything else. As Louis Armstrong said in response to a request to define jazz, "When you got to ask what it is, you never get to know."

The Hard Problem is explaining how subjective experience arises from neural computation. The problem is hard because no one knows what a solution might look like or even whether it is a genuine scientific problem in the first place. And not surprisingly, everyone agrees that the hard problem (if it is a problem) remains a mystery.

Although neither problem has been solved, neuroscientists agree on many features of both of them, and the feature they find least controversial is the one that many people outside the field find the most shocking. Francis Crick called it "the astonishing hypothesis"—the idea that our thoughts, sensations, joys and aches consist entirely of physiological activity in the tissues of the brain. Consciousness does not reside in an ethereal soul that uses the brain like a PDA; consciousness is the activity of the brain.

THE BRAIN AS MACHINE

Scientists have exorcised the ghost from the machine not because they are mechanistic killjoys but because they have amassed evidence that every aspect of consciousness can be tied to the brain. Using functional MRI, cognitive neuroscientists can almost read people's thoughts from the blood flow in their brains. They can tell, for instance, whether a person is thinking about a face or a place or whether a picture the person is looking at is of a bottle or a shoe.

And consciousness can be pushed around by physical manipulations. Electrical stimulation of the brain during surgery can cause a person to have hallucinations that are indistinguishable from reality, such as a song playing in the room or a childhood birthday party. Chemicals that affect the brain, from caffeine and alcohol to Prozac and LSD, can profoundly alter how people think, feel and see. Surgery that severs the corpus callosum, separating the two hemispheres (a treatment for epilepsy), spawns two consciousnesses within the same skull, as if the soul could be cleaved in two with a knife.

And when the physiological activity of the brain ceases, as far as anyone can tell the person's consciousness goes out of existence. Attempts to contact the souls of the dead (a pursuit of serious scientists a century ago) turned up only cheap magic tricks, and near death experiences are not the eyewitness reports of a soul parting company from the body but symptoms of oxygen starvation in the eyes and brain. In September, a team of Swiss neuroscientists reported that they could turn out-of-body experiences on and off by stimulating the part of the brain in which vision and bodily sensations converge.

THE ILLUSION OF CONTROL

Another startling conclusion from the science of consciousness is that the intuitive feeling we have that there's an executive "I" that sits in a control room of our brain, scanning the screens of the senses and pushing the buttons of the muscles, is an illusion. Consciousness turns out to consist of a maelstrom of events distributed across the brain. These events compete for attention, and as one process outshouts the others, the brain rationalizes the outcome after the fact and concocts the impression that a single self was in charge all along.

Take the famous cognitive-dissonance experiments. When an experimenter got people to endure electric shocks in a sham experiment on learning, those who were given a good rationale ("It will help scientists understand learning") rated the shocks as more painful than the ones given a feeble rationale ("We're curious"). Presumably, it's because the second group would have felt foolish to have suffered for no good reason. Yet when these people were asked why they agreed to be shocked, they offered bogus reasons of their own in all sincerity, like "I used to mess around with radios and got used to electric shocks."

It's not only decisions in sketchy circumstances that get rationalized but also the texture of our immediate experience. We all feel we are conscious of a rich and detailed world in front of our eyes. Yet outside the dead center of our gaze, vision is amazingly coarse. Just try holding your hand a few inches from your line of sight and counting your fingers. And if someone removed and reinserted an object every time you blinked (which experimenters can simulate by flashing two pictures in rapid sequence), you would be hard pressed to notice the change. Ordinarily, our eyes flit from place to place, alighting on whichever object needs our attention on a need-to-know basis. This fools us into thinking that wall-to-wall detail was there all along—an example of how we overestimate the scope and power of our own consciousness.

Our authorship of voluntary actions can also be an illusion, the result of noticing a correlation between what we decide and how our bodies move. The psychologist Dan Wegner studied the party game in which a subject is seated in front of a mirror while someone behind him extends his arms under the subject's armpits and moves his arms around, making it look as if the subject is moving his own arms. If the subject hears a tape telling the person behind him how to move (wave, touch the subject's nose and so on), he feels as if he is actually in command of the arms.

The brain's spin doctoring is displayed even more dramatically in neurological conditions in which the healthy parts of the brain explain away the foibles of the damaged parts (which are invisible to the self because they are part of the self). A patient who fails to experience a visceral click of recognition when he sees his wife but who acknowledges that she looks and acts just like her deduces that she is an amazingly well-trained impostor. A patient who believes he is at home and is shown the hospital elevator says without missing a beat, "You wouldn't believe what it cost us to have that installed."

Why does consciousness exist at all, at least in the Easy Problem sense in which some kinds of information are accessible and others hidden? One reason is information overload. Just as a person can be overwhelmed today by the gusher of data coming in from electronic media, decision circuits inside the brain would be swamped if every curlicue and muscle twitch that was registered somewhere in the brain were constantly being delivered to them. Instead, our working memory and spotlight of attention receive executive summaries of the events and states that are most relevant to updating an understanding of the world and figuring out what to do next. The cognitive psychologist Bernard Baars likens consciousness to a global blackboard on which brain processes post their results and monitor the results of the others.

BELIEVING OUR OWN LIES

A second reason that information may be sealed off from consciousness is strategic. Evolutionary biologist Robert Trivers has noted that people have a motive to sell themselves as beneficent, rational, competent agents. The best propagandist is the one who believes his own lies, ensuring that he can't leak his deceit through nervous twitches or self-contradictions. So the brain might have been shaped to keep compromising data away from the conscious processes that govern our interaction with other people. At the same time, it keeps the data around in unconscious processes to prevent the person from getting too far out of touch with reality.

What about the brain itself? You might wonder how scientists could even begin to find the seat of awareness in the cacophony of a hundred billion jabbering neurons. The trick is to see what parts of the brain change when a person's consciousness flips from one experience to another. In one technique, called binocular rivalry, vertical stripes are presented to the left eye, horizontal stripes to the right. The eyes compete for consciousness, and the person sees vertical stripes for a few seconds, then horizontal stripes, and so on.

A low-tech way to experience the effect yourself is to look through a paper tube at a white wall with your right eye and hold your left hand in front of your left eye. After a few seconds, a white hole in your hand should appear, then disappear, then reappear.

Monkeys experience binocular rivalry. They can learn to press a button every time their perception flips, while their brains are impaled with electrodes that record any change in activity. Neuroscientist Nikos Logothetis found that the earliest way stations for visual input in the back of the brain barely budged as the monkeys' consciousness flipped from one state to another. Instead, it was a region that sits further down the information stream and that registers coherent shapes and objects that tracks the monkeys' awareness. Now this doesn't mean that this place on the underside of the brain is the TV screen of consciousness. What it means, according to a theory by Crick and his collaborator Christof Koch, is that consciousness resides only in the "higher" parts of the brain that are connected to circuits for emotion and decision making, just what one would expect from the blackboard metaphor.

WAVES OF BRAIN

Consciousness in the brain can be tracked not just in space but also in time. Neuroscientists have long known that consciousness depends on certain frequencies of oscillation in the electroencephalograph (EEG). These brain waves consist of loops of activation between the cortex (the wrinkled surface of the brain) and the thalamus (the cluster of hubs at the center that serve as input-output relay stations). Large, slow, regular waves signal a coma, anesthesia or a dreamless sleep; smaller, faster, spikier ones correspond to being awake and alert. These waves are not like the useless hum from a noisy appliance but may allow consciousness to do its job in the brain. They may bind the activity in far-flung regions (one for color, another for shape, a third for motion) into a coherent conscious experience, a bit like radio transmitters and receivers tuned to the same frequency. Sure enough, when two patterns compete for awareness in a binocular-rivalry display, the neurons representing the eye that is "winning" the competition oscillate in synchrony, while the ones representing the eye that is suppressed fall out of synch.

So neuroscientists are well on the way to identifying the neural correlates of consciousness, a part of the Easy Problem. But what about explaining how these events actually cause consciousness in the sense of inner experience—the Hard Problem?

TACKLING THE HARD PROBLEM

To appreciate the hardness of the hard problem, consider how you could ever know whether you see colors the same way that I do. Sure, you and I both call grass green, but perhaps you see grass as having the color that I would describe, if I were in your shoes, as purple. Or ponder whether there could be a true zombie—a being who acts just like you or me but in whom there is no self actually feeling anything. This was the crux of a Star Trek plot in which officials wanted to reverse-engineer Lieut. Commander Data, and a furious debate erupted as to whether this was merely dismantling a machine or snuffing out a sentient life.

No one knows what to do with the Hard Problem. Some people may see it as an opening to sneak the soul back in, but this just relabels the mystery of "consciousness" as the mystery of "the soul"—a word game that provides no insight.

Many philosophers, like Daniel Dennett, deny that the Hard Problem exists at all. Speculating about zombies and inverted colors is a waste of time, they say, because nothing could ever settle the issue one way or another. Anything you could do to understand consciousness—like finding out what wavelengths make people see green or how similar they say it is to blue, or what emotions they associate with it—boils down to information processing in the brain and thus gets sucked back into the Easy Problem, leaving nothing else to explain. Most people react to this argument with incredulity because it seems to deny the ultimate undeniable fact: our own experience.

The most popular attitude to the Hard Problem among neuroscientists is that it remains unsolved for now but will eventually succumb to research that chips away at the Easy Problem. Others are skeptical about this cheery optimism because none of the inroads into the Easy Problem brings a solution to the Hard Problem even a bit closer. Identifying awareness with brain physiology, they say, is a kind of "meat chauvinism" that would dogmatically deny consciousness to Lieut. Commander Data just because he doesn't have the soft tissue of a human brain. Identifying it with information processing would go too far in the other direction and grant a simple consciousness to thermostats and calculators—a leap that most people find hard to stomach. Some mavericks, like the mathematician Roger Penrose, suggest the answer might someday be found in quantum mechanics. But to my ear, this amounts to the feeling that quantum mechanics sure is weird, and consciousness sure is weird, so maybe quantum mechanics can explain consciousness.

And then there is the theory put forward by philosopher Colin McGinn that our vertigo when pondering the Hard Problem is itself a quirk of our brains. The brain is a product of evolution, and just as animal brains have their limitations, we have ours. Our brains can't hold a hundred numbers in memory, can't visualize seven-dimensional space and perhaps can't intuitively grasp why neural information processing observed from the outside should give rise to subjective experience on the inside. This is where I place my bet, though I admit that the theory could be demolished when an unborn genius—a Darwin or Einstein of consciousness—comes up with a flabbergasting new idea that suddenly makes it all clear to us.

Whatever the solutions to the Easy and Hard problems turn out to be, few scientists doubt that they will locate consciousness in the activity of the brain. For many nonscientists, this is a terrifying prospect. Not only does it strangle the hope that we might survive the death of our bodies, but it also seems to undermine the notion that we are free agents responsible for our choices—not just in this lifetime but also in a life to come. In his millennial essay "Sorry, but Your Soul Just Died," Tom Wolfe worried that when science has killed the soul, "the lurid carnival that will ensue may make the phrase 'the total eclipse of all values' seem tame."

TOWARD A NEW MORALITY

My own view is that this is backward: the biology of consciousness offers a sounder basis for morality than the unprovable dogma of an immortal soul. It's not just that an understanding of the physiology of consciousness will reduce human suffering through new treatments for pain and depression. That understanding can also force us to recognize the interests of other beings—the core of morality.

As every student in Philosophy 101 learns, nothing can force me to believe that anyone except me is conscious. This power to deny that other people have feelings is not just an academic exercise but an all-too-common vice, as we see in the long history of human cruelty. Yet once we realize that our own consciousness is a product of our brains and that other people have brains like ours, a denial of other people's sentience becomes ludicrous. "Hath not a Jew eyes?" asked Shylock. To-day the question is more pointed: Hath not a Jew—or an Arab, or an African, or a baby, or a dog—a cerebral cortex and a thalamus? The undeniable fact that we are all made of the same neural flesh makes it impossible to deny our common capacity to suffer.

And when you think about it, the doctrine of a life-to-come is not such an up-lifting idea after all because it necessarily devalues life on earth. Just remember the most famous people in recent memory who acted in expectation of a reward in the hereafter: the conspirators who hijacked the airliners on 9/11.

Think, too, about why we sometimes remind ourselves that "life is short." It is an impetus to extend a gesture of affection to a loved one, to bury the hatchet in a pointless dispute, to use time productively rather than squander it. I would argue that nothing gives life more purpose than the realization that every moment of consciousness is a precious and fragile gift.

Brains Wide Shut?[*]

By Patricia Churchland
New Scientist, April 30, 2005

The books-on-consciousness mills are running full tilt. Just about anyone who is conscious seems motivated to write on the subject, and most authors profess themselves emboldened to call their own contribution a theory of consciousness. Because the output is accelerating, it may be innocently assumed that something new has been discovered. Alas, the truth is quite the opposite: very little has been discovered. All this furious activity is reminiscent of the flood of speculative theories of life in the early decades of the 20th century. And for much the same reason: science is moving forward on the problem, but has not yet nailed down the answers—and no one really knows what the answers will look like.

Such a wide-open empirical playing field motivates authors to wrestle feverishly with each other, hurl mud, vote one another off the island, and draw endless boxes with connecting arrows. Naturally, it is vastly easier to hunker down in the hot tub to introspect one's inner milieu than to do the painstaking work of neuroscience. As a result, legions of hopeful Darwins-of-consciousness flock to conferences, vying for attention, and philosophers, horrified by the spectacle of the empirical sciences treading on their sacred territory, try to scare off interlopers with extreme threats of conceptual necessities and logically adequate criteria.

Resolving disagreements about the possible nature of things comes from empirical discoveries about the actual nature of those things. Prior to such resolution, sound and fury emanates from the clash of intuitions and ideologies. And much of it, alas, does signify next to nothing.

A normal part of the maturation of any science, be it physics, astronomy, chemistry, geology or neuroscience, is that "self-evident verities" (otherwise known as "intuitions") expire in the oxygen of factually based theories. Who now bangs his pots concerning the "intrinsic nature" of élan vital (life force)? Or considers it a conceptual necessity that space is Euclidean? And what now of the fixity of continents?

One important lesson that has escaped most philosophers and many neuroscientists is that what seems obvious can change as the science changes. With scientific development comes conceptual development, and this alters how we think about and see the world. This applies to our inner world, as well as to our outer world.

Into this fray rides—once again—the tireless figure of Daniel Dennett, the philosophy professor at Tufts University in Massachusetts whose *Consciousness Explained* brought him such celebrity. Fourteen years on, his new book, *Sweet Dreams: Philosophical Obstacles to a Science of Consciousness* (MIT Press), is a collection of essays devoted mainly to identifying and pummelling those diehard intuitions that he believes, rightly, still obstruct the progress of cognitive neuroscience.

Among these intuitions is the idea that there could be a zombie like me in all respects—all, save that it lacks qualia. A lack of qualia means it doesn't have the "experience" of redness when it sees a London bus, but like me would say: "Look! There is a red London bus." Incredibly (I'm not making this up), zombie-me would have exactly the same conversations about conscious experience that I do. For example, we both say: "When I dream, I am aware of actions, such as flying, but not aware of how bizarre those actions are." The difference is that zombie-me has neither experiences nor qualia to talk about.

Could there be such a zombie? "Perhaps not," says the purveyor of zombies. "It is a thought-experiment-zombie." Fine. But so what? "Well, the mere imagining of such a thing entails that consciousness cannot be a property of the brain..." Good grief. As a colleague once muttered in despair, this argument is not even wrong.

Dennett is right about most of the philosophically pampered intuitions, especially those bravely predicting that "science can never, ever explain consciousness". These intuitions and the arguments they spawn have been repeatedly exposed as confusions, fallacies, circularities, failures of imagination, arguments from ignorance and just plain bunk. By Dennett, to be sure, but by a host of other philosophers as well, including Owen Flanagan and Paul Churchland.

Puzzled by the resistance to his criticism, Dennett has gone further, trying to explain why certain dubious arguments are remarkably resilient in the face of their evident demolition—all to marginal avail, at least for the big money players David Chalmers, Colin McGinn, Ned Block and Frank Jackson.

Why has Dennett's remedial exercise not had greater effect? Part of the answer is old hat: it is the next generation, with less invested in the status quo, on whom the impact of new ideas is usually greatest. But the larger part of the answer, I suspect, is owed to the state of play in neuroscience.

Neuroscience is a very young science, still in search of its own exoskeleton—the fundamental principles that explain how nervous systems work. Although an enormous amount is known about the structure and function of individual neurons, how macro effects emerge from populations of neurons remains poorly understood. For example, we do not yet really understand to what degree sensory or motor systems are hierarchically organised. Deep puzzles endure over the way memories are stored and retrieved, what attention is and how it shifts, and how decisions are made and behaviours organised. And basic issues still need resolving about how neurons code information, and how information is integrated across neuronal populations.

With so many questions so open, neuroscience can't offer an integrated, comprehensive theory of conscious phenomena to oust those old-time intuitions asserting the "metaphysical mysteriousness" of consciousness. Could it do so, some intuitions about, say, the "unity of consciousness" would undoubtedly slink off to join caloric fluid and crystal spheres in the graveyard of scientifically dead ideas.

Dennett himself, though a savvy gamester, may have misplayed his hand in an odd way. In *Consciousness Explained*, the title was not meant to be a joke. Yet those of us who eagerly read the book expecting to understand the neural mechanisms of mental phenomena felt somewhat let down.

Dennett did not explain consciousness in neural terms at all. True enough, he did offer some semifertile new metaphors to substitute for sterile old metaphors, and that was progress of a non-trivial sort. And as in *Sweet Dreams*, he did joyfully beat up some silly ideas. But to describe his story as "explanation" was a bit strong.

What is Dennett's account? First, let's look at two cases where there is a contrast between being aware and not being aware of an event. In the first case, I have been hiking all day, and I have a blister on my heel. I am aware of the pain on my heel and aware of my fatigue. A mother bear and cubs emerge from the bush. Now I am aware neither of my pain nor my fatigue but only of the threat. In the second case, I am in the deep stage of sleep (not dreaming) and you begin to whistle Dixie. I am not aware of hearing the tune.

How does Dennett explain the difference between being aware of my blister before the bear appears and unaware after; and what is the difference between being aware of Dixie when I am awake but unaware while I am in deep sleep?

Dennett argues that at any given moment, many sensory signals, from inside and outside the body, enter the brain. Depending on conditions, some have little effect while others have a big effect. The salience of a signal (the threatening bear) and general state of arousal (being awake) are two relevant conditions among many. When I am aware of the blister, it is because the pain signal hijacks attention, behaviour, memory storage and retrieval, emotions and so on. Enter the mother bear, and the pain loses out to the fear signal, which then dominates the mechanisms for attention, decision making, memory and planning.

Dennett's view is that consciousness of an event is what happens in the brain when a signal takes control of the aforementioned functions: that is, consciousness of an event is a matter of large areas of the brain being influenced by one of a set of competing sensory signals. None of this can happen, he insists, without language. He embraces the surprising conclusion that the non-verbal (animals, infants and the profoundly aphasic) are not genuinely conscious. He has convinced himself that acquiring a language rewires the brain to make it function as a serial digital computer, and that this rewiring is necessary for consciousness. Repeatedly hammered for pampering his own intuition, Dennett doggedly stands by his language requirement for consciousness.

Dennett concedes that he has no understanding of the neural mechanisms of the competition and control dynamics, and cheerfully waves off the brain as merely the wetware implementing the von Neumann-style software. He is strictly a software man, and proud of it.

This is not reassuring. As Dennett must know by now, the hardware-software metaphor applies to brains about as well as it applies to kidneys—poorly—and then only if you blind yourself to glaring disanalogies. Of course it might be convenient if we could understand the brain without neuroscience, but it ain't so.

Quite simply, there is no substitute for understanding the brain at all its levels of organisation. To understand conscious phenomena, we need to understand, in neurobiological terms, such things as the difference between being aware of pain and not being aware of it, why we experience sounds differently from body position or smells, and how autobiographical memories are retrieved.

Apart from lacking "neuro-cred", Dennett's account turns tail when confronted with a range of clinical data from human patients. For example, neuropsychologists David Milner and Melvyn Goodale have carefully studied a patient who, following oxygen deprivation, lost the ability to see shapes. Nevertheless, when placed in front of a box with an oriented slot, and asked to "post" a card in the slot, she is consistently successful in posting the card, regardless of the slot's orientation. The location of her lesion (the ventral stream in the visual pathway) is consistent with a wealth of physiological data identifying the specialised functions in that region.

Why is this a problem for Dennett? Because the shape signal, though not conscious, appears to "take control" to permit her to post successfully. In Dennett's view, consciousness is just winning and taking control. Follow-up psychophysical research on normal subjects indicates that some motor behaviour is routinely guided by non-conscious visual signals that outcompete the conscious visual signals. The Milner and Goodale research is some of the best work on consciousness going.

One rather important upshot of Dennett's claim to have explained consciousness is that the scientifically naive may have been misled into believing that his philosophical story is what an adequate scientific theory of consciousness would look like. Recognition of the inadequacies in Dennett's positive account seemed to invigorate the naysayers' prediction that neuroscientific explanations for conscious phenomena are forever doomed, regardless of what science discovers. The diehards remain faithful to their favourite intuitions, confident that qualia are "ontologically basic" or "transcendentally real"—pick your favourite empty slogan.

So, to avoid equivocation, let's agree that something will be considered a theory of consciousness if, like the theory of how proteins are made, it explains the main properties in sufficient detail to satisfy four conditions: we understand how macro events emerge from the properties and organisation of the micro events; novel phenomena can be predicted; the system can be manipulated; and it is clear at what level of brain organisation the phenomenon resides.

These criteria imply that a genuine explanation of the properties of conscious phenomena must characterise neurobiological mechanisms. A theory satisfying the four desiderata will not be solely a psychological account, linking various cognitive functions, it will not be just an array of boxes, labelled as cognitive, with arrows connecting the boxes. "Boxology" at the psychological level is a crucially important component of this theory, but it does not explain the neurobiological bases for the functions in the boxes. Likewise, an explanatory theory will not consist solely of detailed anatomical maps of what connects to what within the brain, though such maps are also essential to the solution.

And finding correlations for certain conscious events, perhaps via functional MRI or single-cell recordings, does not as such constitute a theory, because such correlations do not, ipso facto, reveal mechanisms. Correlating events using different measures, such as fMRI or behavioural techniques, can be extremely useful, but a roster of correlations doesn't constitute a theory.

So, what can we expect from neuroscience? Most likely, that a theory of consciousness will co-evolve with an understanding of the fundamentals of brain function. And the details? Needless to say, I can't answer that question. The fact is that neuroscience circa 2105 will be profoundly different from neuroscience circa 2005. Even if a kind alien left the answer on my pillow, I could not understand it without also understanding the larger (still missing) theory that embeds it. Likewise, Galileo would have remained stumped about the nature of life had the same alien left him a note describing the structure of DNA, without also teaching him the entire background of modern structural chemistry.

An empirical theory of conscious phenomena will not, or course, simply waft up out of the neural data. It will be the product of brains that create hypotheses, and that creation will draw upon psychology, neuroscience, genetics, computational theory and ethology. Reasonably enough, in *Sweet Dreams* Dennett warns that if you study the growing body of neurobiological data through the lenses of cock-eyed intuitions, then you impede your brain's theory-making machinery.

As we have already seen, his worry is not idle, as some neuroscientists have unwittingly swallowed colourful blather about zombies, "intrinsic properties" and the transcendental mysteriousness of qualia. This book contains many nice remedies against a lot of sneaky rigmarole.

But the unglamorous truth is that science will come to understand the components of consciousness in pretty much the way it has come to understand the components of the nature of life. Not with a single blindingly beautiful insight, but by understanding the mass of detail at many levels of organisation: molecules, cells, networks, subsystems—and the whole system. All of which means a lot of hard empirical slog ahead before we can hand out any Nobels.

Is There Room for the Soul?[*]

New Challenges to Our Most Cherished Beliefs About Self and the Human Spirit

By Jay Tolson
U.S. News & World Report, October 23, 2006

A mind is a tough thing to think about. Consciousness is the defining feature of the human species. But is it possible that it is also no more than an extravagant biological add-on, something not really essential to our survival? That intriguing possibility plays on my mind as I cross the plaza of the Salk Institute for Biological Studies, a breathtaking temple of science perched on a high bluff overlooking the Pacific Ocean in La Jolla, Calif. I have just visited the office of Terry Sejnowski, the director of Salk's Computational Neurobiology Laboratory, whose recent research suggests that our conscious minds play less of a role in making decisions than many people have long assumed. "The dopamine neurons are responsible for telling the rest of the brain what stimuli to pay attention to," Sejnowski says, referring to the cluster of brain cells that produce one of the many chemical elixirs that activate, deactivate, or otherwise alter our mental state. In a deeper way, he explains, evolutionary factors—the need for individual organisms to survive, find food or a mate, and avoid predators—are at work behind the mechanisms of unconscious decision making. "Consciousness explains things that have already been decided for you," Sejnowski says. Asked whether that means that consciousness is only a bit player in the overarching drama of our lives, he admits that it's hard to separate rationalizing from decision making. "But," he adds, "we might overrate the role of our consciousness in making decisions."

Overrated or underrated, consciousness is not being ignored these days. Indeed, during the past 20 years or so it has become the focus of an expanding intellectual industry involving the combined, but not always harmonious, efforts of neuroscientists, cognitive psychologists, artificial intelligence specialists, physicists, and philosophers.

But what, exactly, has this effort accomplished? Has it brought us any closer to understanding how the physical brain is related to the thinking, experiencing, self-aware mind? Is the scientific study of consciousness approaching its own Copernican moment, when the fruits of experimental work yield a compelling, comprehensive theory?

Battle lines. Such questions, and the effort to find their answers, are part of what brought me to La Jolla, home to several prominent centers of consciousness research in addition to Salk. But interesting as the state of the science is, it is not what concerns most owners and users of a mind. There is, indeed, something troubling, if not downright offensive, about the effort to reduce human consciousness to the operations of a 3-pound chunk of wrinkled brain tissue. Such reductionist thinking seems like an assault on the last redoubt of the soul, or, at least, the seat of the irreducible self. Deny or attempt to disprove the immaterial character of the mind, and you elicit some of the same passions that have animated the culture wars over evolution in the classroom, exposing the deep divide between hard-core religious fundamentalists on one side and the equally hard-core scientific fundamentalists on the other.

But if the true believers on both sides of the emerging consciousness debate are likely to shout the loudest on the matter, neither should be allowed to have the last word. There is, in fact, an alternative scenario—one in which the seemingly fixed battle lines of the opposing armies are shown to be drawn according to some rather dubious principles. Not only has advanced neuroscientific research revealed an obdurate mystery at the core of consciousness, but theoretical advances in the natural and physical sciences have greatly complicated the effort to reduce all human phenomena—the mind notably included—to the effects of material causes. And even as cutting-edge science challenges crude materialistic explanations of the phenomenal world, new thinking in philosophy and theology is questioning the assumption of an absolute divide between mind and body, spirit and matter—an assumption that has long sustained many religious conceptions of the soul. Interestingly, these parallel developments in science and religion point to a new picture of reality—or maybe even recall older understandings implicit in traditions as ancient as Judaism or Buddhism—in which subject and object, mind and matter are more interfused than opposed.

Exploring the relationship between the physical brain and consciousness is not simply one of the last great intellectual frontiers. It also sheds light on some of the most vexing life-and-death issues facing us today. The study of consciousness, says Joseph Dial, executive director of the San Antonio-based Mind Science Foundation, which devotes a generous portion of its resources to this field, "has clear clinical applications when you talk about coma and impaired consciousness such as in the Terri Schiavo case. How do you understand consciousness well enough, how do you understand the self and identity well enough, to determine at what point a person is no longer in possession of a self, is no longer conscious in the way we would suggest other humans are conscious and have an identity?"

Consciousness is so tied up with what we think of as our inner selves, our spiritual being, that many of the greatest minds of history have assigned it to an order of reality entirely different from the rest of the natural, physical world. Plato, most influentially, separated the soul, or psyche, from the material body and argued that this reasoning part of our being was immortal. His idea was so powerful and attractive that it has kept philosophers intimately engaged with it to this day. Then, too, because so many influential Christian theologians were part of this philosophical tradition, Platonic ideas have left a lasting imprint on Christian beliefs. The body may die, many Christians hold, but the soul lives on, presumably extending into eternity those qualities that we associate with our conscious minds and our sense of selfhood.

The experimental science that began to emerge in the 17th century would eventually challenge many of the everyday assumptions of the Christian West, including the notion of an Earth-centered cosmos. But few of the great men of early modern science viewed themselves as foes of religion. Few questioned the special status of the soul or its boon companion, the mind. In fact, prominent among the shapers of the scientific worldview was the French mathematician and philosopher Rene Descartes, whose most enduring contribution to modern thought was his argument that reality consisted of two entirely different substances: material substance (res extensa) and thinking substance (res cogitans). But how did these two different substances interact? According to Descartes, the bodily organs sent perceptions and other information via the brain to the mind, located in the pineal gland in the middle of the head. Reflecting upon these data, the mind then made decisions and directed the body's responses, in words or deeds. This dualistic picture of the body-mind relationship would later come to be attacked as the "ghost in the machine" argument. But for centuries, Christians and others found Cartesian dualism a reassuring and reasonable explanation.

Rats and mazes. It would not be long, though, before philosophers and scientists, particularly in the new field of psychology, would turn in earnest to the problem of consciousness, bringing to it not just the experimental methods of investigation but a philosophical conviction that all phenomena were reducible to their more fundamental parts and that the interactions of these parts were governed by discoverable "laws of nature." Following the path of many 19th-century German psychologists, the great Harvard philosopher and scientist William James carried the study of consciousness to impressive lengths, most notably in his 1890 book, *Principles of Psychology*.

But something curious happened within a generation of that book's publication. Psychology quite suddenly dropped the investigation of consciousness. Dissatisfied with the reliance on introspection—how do you make an objective science out of people's subjective reports on their private experiences?—psychologists followed the lead of researchers like Ivan Pavlov and John Watson and turned to the observable results of consciousness: behavior. Or at least most did. For those less enchanted by the business of running rats through mazes there was the siren song of Sigmund Freud's theory of the unconscious mind. For more than half a century, varieties of behaviorism and psychoanalytic theory dominated the field of psychology, banishing the subject of consciousness to the realm of the occult or mere philosophy.

Slowly, however, developments conspired to bring the banished subject back. The invention of program-controlled computers in the 1940s gave birth to artificial intelligence, a branch of computer science dedicated to building machines to accomplish tasks requiring intelligent behavior. At the same time, the effort to create artificial intelligence encouraged a whole new field of psychology concerned with finding universal principles for different mental processes: cognitive psychology.

Also crucial to the rise of the scientific study of consciousness was the very sort of technology so frustratingly unavailable to earlier neuroscientists (including the young Freud), technology that could show what the brain was actually doing when someone experienced the color red or remembered a phone number. A raft of new brain-imaging and scanning technologies, including computed tomography (CT) scans and positron emission tomography (PET) scans, magnetic resonance imaging (MRI) and functional magnetic resonance imaging (fMRI), and magnetoencephalography (MEG), came to fill this need. These instruments enabled researchers to observe brain structure and activity in a variety of noninvasive ways, while the newest gadget in the arsenal, transcranial magnetic stimulation (TMS), actually allows the researcher to disrupt activity in the cortex underlying specific mental tasks. Well before such devices were available, however, in the late 1940s, a Canadian psychologist named Donald Hebb put forth a remarkably resilient hypothesis. According to Hebb, groups of neurons that fire together tend to form what he called "cell-assemblies," the activities of which persist even after the event that triggered their firing is no longer present. These assemblies, in effect, come to represent the triggering event. The neurophysiological basis of thought, Hebb concluded, was the sequential activation of various groups of cell-assemblies.

Variants and refinements of this hypothesis, particularly the notion that neurons that fire together wire together, have been at the center of the research agendas of top cognitive neuroscientists during the past two decades. The most ambitious of these scientists—call them, if you will, the hard-core demystifiers—came to believe quite strongly that most of the mysteries of the mind, if not all of them, are reducible to the biochemical mechanisms underlying these neural networks. These scientists have been a formidable lot, including at least a couple of Nobel laureates who moved to the study of consciousness after doing major work in other fields. One of them, Gerald Edelman, winner of the 1972 prize for his work in immunology, is the founder and director of the Neurosciences Institute, which sits to the west of the Salk Institute on the same La Jolla mesa. Edelman launched the institute in 1981 as part of the Rockefeller Institute in New York City but brought it to La Jolla in 1993, where he also chairs the neurobiology department of the Scripps Research Institute, directly across the street from Neurosciences. A man as conversant with philosophy, literature, and music as he is with science—his early passion was the violin, but he feared he lacked the right stuff to perform—Edelman went into medicine, and then research. As he explains when we meet in his office, Darwin's theory of natural selection is what guided his groundbreaking research on antibody structures, and it is what underlies his theory of neuronal group selection in his work on consciousness. "I wanted to bring Darwin's selectional process to neurons," he says.

Edelman's many books on consciousness explore the various ways that neuronal circuits get established. In the developmental stage of the brain, some neuronal assemblies, or maps, are formed according to genetic rules. Experience then reinforces or weakens these assemblies—or gives rise to new ones—according to how efficiently they respond to signals from the world or the body. The last process, re-entry, is the most difficult to explain, Edelman says, but it is also the most important, since it integrates the activities of various assemblies through what he calls "ongoing parallel signaling between separate brain maps along massively parallel anatomical connections." The binding together of the neuronal activities of maps associated with, say, the perception of an object and those associated with, say, memory, yields an integrated yet highly differentiated experience: a "scene" of primary consciousness that researchers call a quale.

But does the biochemistry underlying these qualia (the plural of quale) adequately account for the experience itself, not to mention aspects of higher-order consciousness that we associate with a sense of self and language? Edelman appears to be of two minds. "We evolved structures that invented language," he says. Yet once humans acquired syntax, Edelman adds, "all bets are off." Biology, he seems to suggest, can take us only so far in understanding the symbol-using mind. "It's not totally reductive," he says. At the same time, among the work being done by the some 36 researchers in Edelman's institute is an ongoing effort to build brain-based devices that perform a task—picking up or avoiding different kinds of objects—not according to an elaborately prescriptive program but by learning from experience, altering, creating, strengthening, and sometimes replacing the synthetic "neural" pathways within its program through success or failure at picking up the right kind of blocks. "Brain-based devices will happen if consciousness is a physical, natural process," Edelman says, clearly implying that it is at least a possibility.

Fuzziness. Another Nobel laureate who turned to consciousness research expressed far less ambivalence about the ability of science to explain the whole mystery. That scientist was Francis Crick, the discoverer, along with James Watson, of the double helical structure of DNA. For roughly two decades after that 1953 breakthrough, Crick helped pioneer molecular and developmental biology. But in 1976, Crick moved from Cambridge University to the Salk Institute to work on a subject that had fascinated him since the early 1950s: the biological basis of consciousness. Not long after, he teamed up with Christof Koch, a promising young German-educated scientist with a degree in physics and an interest in neurons, visual processing, and rock climbing. Together they launched the quest for what they came to call the neural correlates of consciousness, which they defined as "the minimal set of neuronal events that gives rise to a specific aspect of a conscious percept."

The title and first sentence of Crick's 1994 book, *The Astonishing Hypothesis: The Scientific Search for the Soul*, made their ambitious agenda clear: "The Astonishing Hypothesis is that 'You,' your joys and your sorrows, your memories and your ambitions, your sense of personal identity and free will, are in fact no more than the behavior of a vast assembly of nerve cells and their associated molecules."

Although Crick died in 2004, Koch has continued to work on the subject in his laboratory at the California Institute of Technology in Pasadena. In his own book, *The Quest for Consciousness*, he sounds even more confident than his former mentor that focused work on neurons will soon yield not just the correlates but the causes of consciousness. As he demonstrated during the recent 10th annual conference of the Association for the Scientific Study of Consciousness in Oxford, England, Koch can sometimes come across as an affable taskmaster, not quite humorless but still Teutonically firm in telling his colleagues, some 300 cognitive scientists and philosophers on this occasion, where the real investigating should be done.

That domineering tendency surfaces during a debate between him and another prominent consciousness researcher, Susan Greenfield, a professor of pharmacology at Oxford and the director of the Royal Institution of Great Britain in London. Koch, a bit of a showman with his red-dyed hair, yellow shirt, purple tie, and red running shoes, reminds the audience that the great moral of 20th-century biology is specificity, indeed, specific molecular machinery. Dismissing fuzzy holistic approaches, he states his belief that "there are very specific neurons that subserve consciousness." The real challenge, he insists, is to develop genetic techniques to selectively activate and deactivate specific groups of neurons to see how they are related to different conscious states.

Greenfield, no shrinking violet herself, makes it clear that she finds Koch's agenda much too restrictive. She is interested in the broader problem of the gap between consciousness and unconsciousness, or, really, the continuum between the two states. "I suggest that consciousness is continuously variable," she says, "that there are varying degrees of consciousness." Greenfield emphasizes the importance of the neuronal assembly—the nets of neurons that extend over wide areas of the brain—and particularly the neuromodulating chemistry that activates these assemblies, bringing them into concerted focus for less than a second, until they are supplanted by the activation of other (possibly closely related or associated) neuronal assemblies. The neuromodulators are the underlying chemistry of mood, emotions, and feelings, and, as Greenfield has written in her book, *The Private Life of the Brain*, "emotions are the most basic form of consciousness."

It is at moments like this that the definitional fuzziness of the enterprise can hit you full force. Are the two debaters really talking past each other? Are they even talking about the same thing? David Galin, a neuropsychiatrist and professor emeritus at the University of California-San Francisco, makes the point that researchers are often in such a hurry to explain consciousness in terms of their pet theories that they don't adequately examine just what they are trying to explain. "People treat consciousness as a thing," he says, "or as the system that generates the qualia or as the central mechanism that directs how it is employed—and those are all different things."

States of experience. At least some of the philosophers involved with the modern study of consciousness have been aware of this problem from the earliest years of the enterprise. One of them, David Chalmers, now at the Australian National University, is widely known within the field for a speech he gave in 1994 at the first of a series of ongoing biannual conferences on consciousness held in Tucson, Ariz. (Out of these conferences would come both *The Journal of Consciousness Studies* and the Center for Consciousness Studies at the University of Arizona, a center that Chalmers would come to direct in 1997 and that is now directed by Stuart Hameroff, a professor of anesthesiology at the same university.)

Chalmers created a stir at Tucson I by trying to clarify the "hard problem" of consciousness: the problem, as he put it, of experience itself. "When we think and perceive there is a whir of information-processing," Chalmers declared, "but there is also a subjective aspect." This aspect, he continued, "is experience. When we see, for example, we experience visual sensations: the felt quality of redness, the experience of dark and light, the quality of depth in a visual field . . . . Then there are bodily sensations, from pains to orgasms; mental images that are conjured up internally; the felt quality of emotion, and the experience of a stream of conscious thought. What unites all of these states is that there is something it is like to be in them. All of them are states of experience."

The easier questions for Chalmers were things like the ability to discriminate among assorted stimuli or to report upon mental states. But subjectivity arising out of matter: that to Chalmers was a mystery so seemingly insoluble that he wrote a whole book (*The Conscious Mind: In Search of a Fundamental Theory*) arguing that consciousness had to be considered a fundamental category like space, time, or gravity—explicable only by special, psychophysical laws.

Some cognitive theorists, including Tufts University philosopher Daniel Dennett, have accused Chalmers of making many difficult but surmountable problems into one mighty, insurmountable one. Explain all the little problems, Dennett insists, and you solve the big one—or dissolve it. Dennett is a genial figure, but he can be a bulldog of physical reductivism, quick to sniff out and attack anyone he thinks might be sneaking back to Cartesian dualism. He also enjoys a certain enfant terrible status for his most recent book, *Breaking the Spell: Religion as Natural Phenomenon*, which enrages many believers with its Darwinian dissection of the religious impulse.

Dennett exudes confidence in his own position: that consciousness is about "fame in the brain," to use his now famous phrase. At any one moment, Dennett argues, there are many potential conscious states, many contending neuronal assemblies, vying for celebrity, their big moment under the lights. But only one of these "multiple drafts" wins the competition, perhaps selected by the kind of Darwinian survival-enhancing mechanisms that Salk's Sejnowski and others study. The big mistake, according to Dennett, is to think that there is some homunculus of a self sitting in the theater of the brain and observing, or even directing, the ongoing show. "This is our old nemesis, the Audience in the Cartesian Theater," Dennett wrote in his 1991 book, *Consciousness Explained*.

When I ask Dennett if he feels that his ideas have been vindicated by research during the past 15 years, he answers in the unwavering affirmative: "The idea of fame in the brain and parallel competition seems to be an idea that works pretty well," he says. "Now we can begin to talk about what the conditions of the competition are, where they occur, why and how they occur."

The more you talk to Dennett, though, the more you sense that what he is really interested in, once all the neurophysiological conditions of the competition have been worked out and explained, is higher-order consciousness. "Language changes everything," Dennett says, sounding a lot like Edelman. But when I ask whether that means that meaning is created by symbol-wielding consciousness, Dennett insists that it does not. "This is what I've meant over the years when I've said that the brain is a syntactic engine mimicking a semantic engine." By that, Dennett presumably means that consciousness produces orderly, grammatical representations of something out there in the world that is meaningful, but it does not create meaning. It is not necessary to meaning.

Which of course raises the question of what Dennett means by meaning. He explains by describing his fundamental disagreement with another leading philosopher of consciousness, John Searle, author of *The Rediscovery of Mind*: "Once we understand how there can be a machine that tracks meaning, an organism that tracks meaning," says Dennett, "then we can start asking what more is special about consciousness. This is exactly the other way around from, say, John Searle, who says there is no meaning without consciousness, that we have to do consciousness first, and that nothing can mean anything if it weren't for consciousness. I say, 'Oh no, on the contrary, there is meaning in microorganisms where there is no consciousness, because it's the appropriate response to information in the service of life—that's where meaning comes in.'"

Survival machines. If that's what meaning fundamentally comes down to— the sum of appropriate responses to information in service to life—it is easy to see why so many people view the study of consciousness as a potentially dispiriting project. If consciousness, particularly higher-order consciousness, exists only to respond more effectively to information in service to life, then we are nothing more than Darwinian survival machines. Other notions of value, purpose, freedom, and individuality—notions as important to many secular humanists as to religious people—are reduced to, at best, reassuring illusions of possible survival value. Other, more religiously grounded notions of spirit and soul get even shorter shrift in this reductionist view.

But need the findings and insights of the study of consciousness lead to such a dispiriting conclusion? For two reasons, it would seem not. One reason lies in science itself—specifically, in a sophisticated critique of the reductive materialism that came to dominate modern experimental science during its so-called classical phase from the 17th century to the early 20th century. That critique emerges from frontier work in many areas, particularly physics, suggesting that the search for ultimate causality in smaller and smaller bits of matter is finally a bootless enterprise. The further one goes down the scale of physical reality, the less material matter appears to be. In fact, the further one goes down, the more reality seems to consist of nonmaterial information, pure potentialities of matter or energy but not quite either. Quantum mechanics has demonstrated the flux of particle and wave at subatomic levels, suggesting that the only fixity at such levels comes from the act of observing the object and arresting it at one or another stage of its being.

This point about the role of the observer raises particularly interesting questions about the power of human consciousness not just to define but to influence physical reality (including the physical brain), a point that has been explored by, among others, Henry Stapp, a physicist at Lawrence Berkeley National Laboratory in Berkeley, Calif. His argument, elaborated in his book *Mind, Matter and Quantum Mechanics*, proposes that conscious experience is not a mere product of underlying brain activity but an interactive event in which the attention and intention of the observing mind also have effects on the brain. To some biological reductionists, this notion of top-down (or mind-brain) effects is heresy, but its intellectual appeal reaches well beyond quantum physicists.

Not only quantum mechanics but a number of new fields such as the science of complexity put into question the whole enterprise of explaining reality in terms of bottom-up causality alone. As Galin points out, that kind of thinking only reversed the old, prescientific hierarchical conception of top-down causality, an explanation that attributed ultimate causality to a divine being or prime mover. In thinking about a phenomenon like consciousness, many today argue that it might be useful to move beyond the hierarchical model of causality and consider whether causality moves in both directions, up and down, between different levels of complex systems or organizations. It might be useful also to think of the mind as what philosopher Philip Clayton, a professor at Claremont Graduate University, calls an emergent property, a complex system that is more than the sum of its parts and that has effects on the systems that support it. One of the things that distinguish the "moreness" of mind, according to Clayton (and Stapp would agree), is its unique ability to represent, know, and interpret the objects of its own awareness, an ability that makes it possible for a human being to make decisions and initiate actions and not just to be acted upon, or determined, by a lengthy chain of survival-related factors. This is not to say that the mind is not strongly concerned with, or shaped by, the exigencies of survival. But for Clayton, the mind is more than the sum of the parts that support it because it is a semantic machine and not just the elaborately embodied computer, or syntactical machine, that Dennett says it is. It is not, in other words, a machine that merely responds to external stimuli or underlying physical factors that subserve it. Mind—at least higher-order consciousness—is, by this reasoning, very much involved in creating meaning, largely if not entirely through its ability to assert the existence of things through language.

If the fundamental levels of reality are more informational than material, as quantum physics suggests, then consciousness may be the interface between the fundamental quantum world of information and the "classical" physical world that is more accessible to our senses. That, at least, is a theory developed by Oxford physicist Roger Penrose and Hameroff. Penrose came first to this idea while wrestling with the problem of how we understand mathematics if understanding is not just following a rule (in the way a computer does) but requires understanding the meaning of mathematical concepts. To answer this, Penrose proposed that consciousness was a quantum computation within the brain, an infinitesimal collapse of quantum information into classical information that takes place at the level of the neurons. Impressed by Penrose's argument, Hameroff approached him with the suggestion that the site of this collapse might be at the more microscopic level of the microtubule, a computerlike protein structure inside the dendrites of every neuron and, indeed, every cell.

Hard-line. Although the theory is very far from being proved—and many neuroscientists, including Koch, scoff at it as being completely untestable—Hameroff has published a list of 20 testable predictions, and he claims that some have been confirmed. More broadly, though, the line of inquiry that Penrose and Hameroff have opened, and which has been differently explored by other physicists like Stapp, suggests that consciousness is far more than a sophisticated survival machine or even a highly agile embodied computer. Instead, the mind's resistance to simple reductive explanations lends support to the notion that it is a profoundly complex emergent system whose capacity for intentional acts and creative discoveries connects it with the underlying order of reality, an order analogous, Hameroff suggests, to the world of forms or ideas that Plato believed stood behind our shadowy and ephemeral world of appearances.

Within religion itself there is also fresh thought about the implications of the new science of the mind for core religious principles and beliefs. Malcolm Jeeves, an honorary professor of psychology at the University of St. Andrews, is one of many believing scientists who think the Christian concept of the soul should be relieved of its Cartesian and Platonic overlays. "The immortality of the soul is so often talked about that it is easy to miss that the Jewish view did not support it," Jeeves says. "Furthermore, the original Christian view was not the immortality of the soul but the resurrection of the body." But Platonism did creep in, Jeeves acknowledges, winning over such influential Christian theologians as Augustine and John Calvin. In Jeeves's view, the new science of consciousness, by showing the inseparable links between mind and body, restores the original Christian conception of the unity of the person. As many Christian theologians now say, human beings do not have souls; they are souls. But Jeeves is realistic in thinking that it will take decades for many of his fellow Christians to accept this way of viewing the soul. And that acceptance will not be made easier by the hard-line reductivism of people like Dennett and Crick who, Jeeves says, "commit the fundamental error of nothing-buttery."

But grant Dennett and many other cognitive scientists their view that the self is not a spectator in the theater of consciousness but the composite of multiple drafts related to and constituting the biography of that particular individual. If this view is true, where is the self or identity on which even a broad-minded religious believer might base his notion of the soul?

Here Christians and others might turn to the wisdom of Buddhism, in which the self is correctly understood not as an entity or substance but as a dynamic process. As Galin writes in a collection of essays on Buddhism and science, this process is "a shifting web of relations among evanescent aspects of the person such as perceptions, ideas, and desires. The Self is only misperceived as a fixed entity because of the distortions of the human point of view." The Buddhist concept of *anatman* does not suggest that the self is nonexistent but rather asserts that it cannot be reduced to an essence.

Galin proposes that rehabilitating the notion of spirit may be the best way to a new understanding of the self in a post-dualist age. The experience of spirit, he argues, is itself part of the human capacity to experience implicit organization, hidden order, deeper and ineffable connectedness in what we see or otherwise encounter, whether a magnificent work of architecture like Notre Dame or a spectacular vista such as the Grand Canyon. Experiencing spirit is finding unity and wholeness in something, and Galin suggests that we view the self as spirit in that sense: the organization—or even the emergent property—of all of a person's subsystems, not just one more subsystem.

In recent years, the scientific study of consciousness has taken bold, if not always steady, steps in the direction of understanding the experience of wholeness and human spirituality in general. One prominent researcher, Andrew Newberg, a professor of nuclear medicine at the University of Pennsylvania, directs his university's recently founded Center for Spirituality and the Mind, a cross-disciplinary program devoted in part to the fledgling field of "neurotheology." In one respect, this venture marks yet another return to the legacy of William James, whose later work included his masterful *Varieties of Religious Experience*. The findings of Newberg and his late colleague, Eugene D'Aquili, do not yet rise to the Jamesean level, but they do point in a promising direction. They even suggest that if religion can learn something valuable about the unity of body and mind from science, then science might be able to relearn something from religion about the deepest purposes of our minds.

Consciousness in the Raw[*]

The Brain Stem May Orchestrate the Basics of Awareness

By Bruce Bower
Science News, September 15, 2007

In October 2004, Swedish neuroscientist Bjorn Merker packed up his video camera and joined five families for a 1-week get-together in Florida that featured several visits to the garden of childhood delights known as Disney World. For Merker, though, the trip wasn't a vacation. With the parents' permission, he came to observe and document the behavior of one child in each family who had been born missing roughly 80 percent of his or her brain.

These children, 1 to 5 years old at the time of their Disney adventure, had suffered strokes as fetuses or had experienced other medical problems shortly before or after birth that destroyed nearly all of the brain's outer layer, or cortex. In this rare condition, called hydranencephaly, cerebrospinal fluid fills the gaping hole within the child's head.

Such youngsters often die in the first year of life as a result of seizures, cerebral palsy, lung abnormalities, and a variety of other physical ailments. With proper medication and the installation of shunts to drain fluid from the braincase, however, some individuals live 20 years or more.

Neurologists typically regard hydranencephaly as an anatomical sentence to a lifelong "vegetative state." Such children supposedly validate a brutally simple equation: Little or no cortex equals no awareness of any kind.

In family activities observed in the Magic Kingdom and elsewhere, the kids quickly cast doubt on that standard assumption. Merker noted that these cortex-deprived, nonverbal children remained alert for much of the day. They reacted to what happened around them and expressed a palette of emotions. A 3-year-old girl's mouth opened wide and her face glowed with a mix of joy and excitement when her parents placed her baby brother in her arms.

The youngsters displayed good hearing but limited eyesight, a curious pattern given that they typically retained small parts of the visual cortex but none of the auditory cortex.

In observations at each child's home, Merker noted that these youngsters recognized familiar adults, liked familiar settings, and preferred specific toys, tunes, or video programs. Although saddled with limited mobility, some kids took behavioral initiatives, such as learning to activate a toy by throwing a switch.

In the February *Behavioral* and *Brain Sciences*, Merker, an independent neuroscientist in Segeltorp, Sweden, described how the accomplishments of these children relate to behaviors recorded in prior studies of human-brain function and of animals after surgical removal of the cortex. His analysis generates a provocative proposal: Basic awareness of one's internal and external world depends on the brain stem, the often-overlooked cylinder of tissue situated between the spinal cord and the cortex.

Merker argues that the brain stem supports an elementary form of conscious thought in kids with hydranencephaly. It also contains auditory structures capable of preserving hearing in someone without a cortex. In contrast, optic nerve damage in hydranencephaly frequently impairs vision, regardless of what the brain stem does.

Self-awareness and other "higher" forms of thought may require cortical contributions. But Merker posits that "primary consciousness," which he regards as an ability to integrate sensations from the environment with one's immediate goals and feelings in order to guide behavior, springs from the brain stem.

If he's right, virtually all vertebrates—which share a similar brain stem design—belong to the "primary consciousness" club. Moreover, medical definitions of brain death as a lack of cortical activity would face a serious challenge. At the very least, physicians could no longer assume that individuals with hydranencephaly don't need pain medication or anesthesia during invasive medical procedures.

"To be conscious is not necessarily to be self-conscious," Merker says. "The tacit consensus concerning the cerebral cortex as the 'organ of consciousness' . . . may in fact be seriously in error."

BRAIN DRAIN

The roots of Merker's thesis emerged more than 50 years ago in the operating room of Canadian neurosurgeons Wilder Penfield and Herbert Jasper. The surgeons pioneered the removal of large chunks of cortex as a treatment for severe, uncontrolled epilepsy. To identify and avoid damaging still-functional brain areas, Penfield and Jasper kept patients awake during the surgery and administered only local anesthesia.

Various mental abilities suffered during and after the operations, depending on the site and extent of the neural loss. Nevertheless, patients maintained a conscious stream of thought, Penfield and Jasper found.

In the course of electrically stimulating various brain areas during operations to identify key functional areas, they noted that current delivered to the right spots could produce every kind of seizure except one—so-called "absence epilepsy" characterized by a sudden loss of consciousness for a few seconds. On the basis of what they knew about the brain, the researchers theorized that structures within and just above the brain stem typically trigger absence epilepsy and collaborated with the cortex to regulate conscious thought and intentional acts.

Animal research, predominantly with rats, has since indicated that three adjacent parts of the brain stem comprise a "neural reality simulator" that gives rise to a fundamental form of consciousness, Merker asserts.

Along the top of the midbrain, which represents the roof of the brain stem, layers of cells interpret the spatial layout of an animal's surroundings relative to its body. Just below, a patch of gray tissue influences emotion-related behaviors, such as aggression, sex, defensive maneuvers, and pain reactions.

Farther down the brain stem lie interconnected regions that regulate the direction of eye gaze and organize decisions about what to do next, such as reaching for a piece of food or pursuing a potential mate.

Together, these structures surround brain stem tissue that connects to sensory areas throughout the cortex.

Merker proposes that, in creatures with a brain stem but little cortex, the neural reality simulator produces a two-dimensional, screenlike map of the world featuring moving shapes. A large cortical endowment beefs up the neural reality simulator, creating an ability to perceive a three-dimensional world composed of solid objects. Neural expansion also allows people to reflect about what they think and feel.

In support of his theory, Merker cites studies conducted over the past 40 years in which rats and cats showed relatively few behavioral problems after surgical removal of the cortex, either in infancy or adulthood. These cortex-deprived animals use vision and touch to orient to their surroundings, learn where to find food in mazes, and remain capable of standing, climbing, grooming, mating, and caring for offspring.

Merker also cites an unusual phenomenon known as the Sprague effect. Complete removal of the visual cortex on one side of the brain renders animals unable to see anything in the half of the visual field opposite the surgical site. Yet a tiny cut to the midbrain restores the animal's ability to detect and approach moving entities, even though it still can't distinguish one object from another.

The Sprague effect underscores the brain stem's visual influence, Merker argues. Visual-cortex removal details brain stem activity via numerous neural links to the midbrain's spatial cells that suddenly lack meaningful input. A well-placed midbrain cut halts activity by some of those wayward connections, allowing a partial return of sight, in his view.

Any entity with the equivalent of a neural reality simulator, "whether cast in a neural medium or eventually in silicon," would experience consciousness, Merker theorizes.

CORTICAL DIVIDE

Of 27 comments by mind and brain researchers published with Merker's article, nearly half agreed that the inner workings of consciousness lie in the brain stem.

"The roots of consciousness exist in ancient neural territories we share with all vertebrates," says neuroscientist Jaak Panksepp of Washington State University in Pullman. "By the weight of empirical evidence, all mammals are sentient beings."

In his own research, Panksepp studies the ability of animals to experience biologically based states of mind or feelings that range from hunger and thirst to emotional delight and distress. For instance, Panksepp and a coworker reported in a controversial 2003 paper that rats express "joy" while playing with other rats by making ultrasonic sounds that represent an ancestral form of laughter.

Psychologist Carroll Izard of the University of Delaware in Newark emphasizes that this form of primary consciousness, as Merker would put it, or "primary affect," as Panksepp terms the rats' consciousness, consists of sensory activity in the brain stem. This capacity generates emotions and an awareness of one's surroundings but not an ability to talk about what one has experienced, Izard continues. In the same way, people can become conscious of a feeling that they can't label or describe, a phenomenon that's especially common in healthy infants and in children lacking a cortex, Izard says.

The existence of primary consciousness challenges widespread assumptions among physicians that newborns and fetuses can't feel pain, adds pediatric neurologist K.J.S. Anand of the University of Arkansas for Medical Sciences in Little Rock. Evidence now suggests that adult and immature brains use different systems to process pain, Anand says.

The brain stem and the thalamus, a relay station for sensation just above the brain stem, foster pain responses in babies before and after birth, he asserts. The cortex takes over pain perception as it greatly expands during childhood and adolescence, Anand hypothesizes.

Other investigators criticize Merker for denying the cortex its traditional position as the brain's engine of consciousness. Even if a basic form of consciousness exists, they regard it as at least a partial product of the cortex, not just the brain stem as Merker argues.

Conscious thought probably relies on the workings of connected brain areas within and outside the cortex, contend Susanne Watkins and Geraint Rees, neuroscientists at University College London. "It seems unlikely that activity in any single area of the human brain will be sufficient for consciousness," they write.

Children with hydranencephaly studied by Merker possess remnants of cortical tissue that could have triggered states of awareness, the researchers suggest

Other commenters, including philosopher Gualtiero Piccinini of the University of Missouri-St. Louis, cite prior evidence that the cortex by itself regulates visual awareness. Following visual-cortex damage, certain patients report no conscious ability to see on one side of their visual field but still unconsciously perceive the identity and location of items in that same visual field. Scientists call this phenomenon Blindsight.

The most extensively studied Blindsight patient has frequently reported being aware of "something" in his blind visual field, Merker notes. This man retains primary visual consciousness of his surroundings but can't describe what he sees in words, the Swedish researcher contends.

RECLAIMED KIDS

In a 1999 report, D. Alan Shewmon, a pediatric neurologist at the University of California, Los Angeles Medical Center, and his colleagues described home observations of three children, ages 6 to 17, who had been born with hydranencephaly and raised by loving, attentive parents. Each child displayed comparable signs of conscious mental activity, the researchers reported.

For instance, shortly after birth, a newborn girl's brain scan revealed an almost total lack of cortical tissue. Physicians told the girl's mother that the child would live no more than 2 years as a "vegetable." A neurologist concluded that the girl's brain was "like that of a reptile" and that she would never interact with other people.

Shewmon first visited the girl at age 5, observing her behavior at home. Despite difficulty sitting up or walking without aid, she exhibited excellent health. She smiled in response to Shewmon's friendly overtures and immediately looked at objects brought close to her. In a videotaped play session with her mother, the girl uttered "ah-ah" when encouraged to say "mama."

She brightened upon hearing happy songs, but often cried during sad songs. She enjoyed the sensory stimulation of car rides, crying at stops and calming down as motion resumed. She disliked the loud noises of vacuum cleaners and hair dryers. She demonstrated understanding of a few words, including "bunny rabbit" for one of her stuffed toys.

"If these children had been kept in institutions or treated at home as 'vegetables,' there can be little doubt that they would have turned out exactly as predicted," Shewmon says.

After making his own observations of children with hydranencephaly and their families, Merker seconds that point. He notes that well-treated youngsters born with little or no cortex regularly display brief losses of consciousness due to absence epilepsy, a clear sign that at other times they're conscious.

Parents described these lapses of awareness in their children to Merker with phrases such as "she is off talking with the angels."

Perhaps most intriguingly, kids with hydranencephaly demonstrate that the brain stem is not simply a reptilian relic stashed in the brain's basement. "The human brain stem is specifically human," Merker says. "These children smile and laugh in the specifically human manner, which is different from that of our closest relatives among the apes."

For now, the neural puzzle of consciousness remains unsolved. But cortically endowed investigators may have much to learn from cortically deprived kids.

Bibliography

Books

Amen, Daniel G. *Magnificent Mind at Any Age: Natural Ways to Unleash Your Brain's Maximum Potential*. New York: Harmony, 2008.

Cohen, Gene D. *The Mature Mind*. New York: Basic Books, 2005.

Cozolino, Louis. *The Healthy Aging Brain: Sustaining Attachment, Attaining Wisdom*. New York: W.W. Norton & Co., 2008.

Damasio, Antonio R. *Descartes' Error*. New York: Putnam 1994.

Deacon, Terrence W. *The Symbolic Species: The Co-Evolution of Language and the Brain*. New York: W. W. Norton & Company, 1997.

Dennett, Daniel C. *Consciousness Explained*. Boston: Little Brown & Co., 1991.

Doidge, Norman. *The Brain That Changes Itself: Stories of Personal Triumph from the Frontiers of Brain Science*. New York: Viking Adult, 2007.

Edelman, Gerald M. *Second Nature: Brain Science and Human Knowledge*. New Haven: Yale University Press, 2006.

Gazzaniga, Michael S. *Human: The Science Behind What Makes Us Unique*. New York: Ecco, 2008.

Feinberg, Todd E. *Altered Egos: How the Brain Creates the Self*. Oxford University Press, 2001.

Fortanasce, Vincent. *The Anti-Alzheimer's Prescription: The Science-Proven Plan to Start at Any Age*. New York: Gotham, 2008.

Greenfield, Susan A. *The Human Brain: A Guided Tour*. New York: Basic Books, 1997.

Humphrey, Nicholas. *Seeing Red: A Study in Consciousness (Mind/Brain/Behavior Initiative)*. Cambridge, Mass.: Harvard University Press, 2006.

Koch, Christof. *The Quest for Consciousness: A Neurobiological Approach.* Denver, Colo.: Roberts & Company Publishers, 2004.

Linden, David J. *The Accidental Mind: How Brain Evolution Has Given Us Love, Memory, Dreams, and God.* Cambridge, Mass.: Belknap Press of Harvard University Press, 2007.

Lynch, Gary, and Richard Granger. *Big Brain.* New York: Palgrave Macmillan, 2008.

Marcus, Gary Kluge: *The Haphazard Construction of the Human Mind.* Boston: Houghton Mifflin Co., 2008.

Pinker, Steven. *How the Mind Works.* New York: Norton, 1997.

Ramachandran, V.S., and Sandra Blakeslee. *Phantoms in the Brain: Probing the Mysteries of the Human Mind.* New York: William Morrow and Company, Inc., 1998.

Sabbagh, Marwan. *The Alzheimer's Answer: Reduce Your Risk and Keep Your Brain Healthy.* Hoboken, N.J.: John Wiley & Sons, 2008.

Sacks, Oliver. *The Man Who Mistook His Wife For A Hat: And Other Clinical Tales.* New York: Summit Books, 1985.

Schwartz, Jeffrey M., and Sharon Begley, *The Mind and the Brain: Neuroplasticity and the Power of Mental Force.* New York: Regan Books/HarperCollins, 2002.

Sternberg, Eliezer J. *Are You a Machine?: The Brain, the Mind, And What It Means to Be Human.* Amherst, N.Y.: Humanity Books, 2007.

Taylor, Jill Bolte. *My Stroke of Insight: A Brain Scientist's Personal Journey.* New York: Viking Adult, 2008.

Trimble, Michael R. *The Soul in the Brain: The Cerebral Basis of Language, Art, and Belief.* Baltimore: The Johns Hopkins University Press, 2007.

Web sites

Readers seeking additional information on the brain may wish to consult the following Web sites, all of which were operational as of this writing.

Alzheimer's Association

www.alz.org

Alzheimer's Association attempts to fight Alzheimer's disease by promoting research, providing support for those afflicted, and spreading information on how to maintain a healthy brain.

Project on the Decade of the Brain

www.loc.gov/loc/brain/

President George H. W. Bush proclaimed the 1990s "The Decade of the Brain," and this site details efforts by the Library of Congress and the National Institute of Mental Health of the National Institutes of Health to raise awareness of the many breakthroughs in brain research then being made.

Additional Periodical Articles with Abstracts

More information about the brain and related subjects can be found in the following articles. Readers interested in additional articles may consult the *Readers' Guide to Periodical Literature* and other H.W. Wilson publications.

A Healthy Obsession. Sarah A. Klein. *Bicycling* v. 49 pp. 23-4 November 2008

It is possible to become addicted to bicycling. Vigorous exercise alters brain chemistry, according to John J. Ratey, the author of *Spark: The Revolutionary New Science of Exercise and the Brain* and an associate clinical professor of psychiatry at Harvard Medical School. Exercise increases the body's levels of serotonin, dopamine, and norepinephrine, chemicals that enable neurons to communicate with one another. Scientists have long suspected that dopamine, in particular, plays a significant role in fueling human drive. The theory is that the brain releases dopamine to prompt behavior linked with survival, such as eating, having sex, and winning money. Dopamine makes these experiences more salient and memorable—and more enjoyable, which motivates people to repeat them. As a result, Ratey says, it is possible to become addicted to the chemical changes that exercise generates.

Who Are You? *Current Health 2* v. 34 p5 November 2007

Some people find it very difficult to recognize people they know. They suffer from a condition called prosopagnosia, or face blindness. Researchers believe that it is associated with the part of the brain—the fusiform gyrus—that responds to faces. In an interview, Richard Russell, a scientist at the Prosopagnosia Research Center at Harvard University, explains the disease.

The Discover Interview: Steven Pinker. Marion Long. *Discover* v. 28 pp48-50, 52, 71 September 2007

An interview with renowned cognitive scientist Steven Pinker. For more than 25 years, Pinker has been a driving force in linguistics theory, analyzing language in laboratories at Massachusetts Institute of Technology; Stanford University, California; and Harvard University, Massachusetts, where he is currently the Johnstone Family Professor of Psychology. As communicated in his popular books, Pinker has developed a theory of the evolution of the mind and the source of language that argues that the brain comes preprogrammed with many behavioral dispositions and talents. In the interview, Pinker discusses his linguistic theories as well as his book *The Stuff of Thought: Language as a Window Into Human Nature*.

Rise of the Cyborgs. Sherry Baker. *Discover* v. 29 pp50-2, 54-7 October 2008

Part of a special section on discoveries that could change the world. Researchers are investigating the potential of brain-computer interfaces. Brain-computer interfaces decode the conscious intentions carried by an individual's neural signals. Current work targets the severely disabled and patients with locked-in syndrome who are unable to speak or communicate their needs, with the interface enabling them to spell out words on a computer screen. Based on some studies, researchers envisage brain-computer interfaces that will give such individuals an actual voice and enable paralyzed individuals to walk, reach, and grasp. Some investigators also think that brain-computer interfaces will be developed that will endow humans with extraordinary powers, such as interspecies communication. A sidebar describes how scientists map the brain activity of patients.

Is There an Inner Zombie Controlling Your Brain? Carl Zimmer. *Discover* v. 29 p80 October 2008

Research conducted over the last 40 years suggests that an unconscious part of the brain controls much of what an individual thinks and does. From the late 1960s onwards, psychologists and neurologists began to find evidence that individuals are strongly influenced by perceptions, thoughts, feelings, and desires of which they are totally unaware. Given increasing evidence for unconscious control in the brain, some scientists have began to denigrate the importance of self-awareness. Experiments of conscious and unconscious processing of information, however, have shown that, although some sorts of tasks may be carried out in the absence of any consciousness, others require consciousness and self-awareness.

Rewiring the Creative Mind. Gregory Berns. *Fast Company* p51, 54 October 2008

Recent advances in neuroscience, prompted by functional magnetic resonance imaging that allows scientists to monitor brain activity better than ever, have changed what we know about key attributes of creativity. These advances, for example, have shown that how people perceive something is not merely a product of what eyes and ears transmit to the brain but is a product of the brain itself. The significance of these findings about perception for iconoclasts is discussed.

Hands Up If You Think You've Got Free Will. Chris Frith. *New Scientist* v. 195 pp46-7 August 11, 2007

Scientists are rather a long way from proving that free will is not an illusion. In an experiment to assess free will, Benjamin Libet at the University of California, San Francisco, found that a change in his subjects' brain activity occurred 300 milliseconds before they reported the urge to lift their fingers, showing that it is the brain and not free will that causes actions. The implication that humans do not have conscious control and that actions are predetermined by the genetic makeup and environmental history embodied in a person's brain worries philosophers and theologians because it questions the basis of morality. The sense of people being responsible for their actions is also crucial for the smooth functioning of society. However, for scientists to prove the existence of free will, they will have to solve the hard problem of exactly how a desire in the mental realm can cross into the physical world and cause something to happen.

Mind Reading Is Now Possible. Sharon Begley. *Newsweek* v. 151 p22 January 21, 2008

The possibility of reading thoughts by decoding brain-activity patterns is becoming increasingly possible. According to John-Dylan Haynes of the Max Planck Institute for Human Cognitive and Brain Sciences in Leipzig, Germany, every thought is associated with a pattern of brain activity, and a computer can be programmed to recognize the pattern associated with a particular thought. Scientists at Carnegie Mellon University are researching a method using functional magnetic resonance imaging technology that appears to have the capacity to isolate and identify specific thought patterns. Moreover, the brain-activity patterns are eerily consistent from one person to another, raising the possibility of a universal mind-reading dictionary, in which a specific brain-activity pattern is related to a specific thought in most people. It seems inevitable that mind reading will be used in criminal and terrorism cases, although this raises issues of reliability and self-incrimination.

Sad Brain, Happy Brain. Michael Miller, Michael Craig. *Newsweek* v. 152 pp50-2, 56 September, 22 2008

Scientists today are finding out more about the fascinating biology behind people's most complex feelings. The brain is responsible for most of what one cares about—language, creativity, imagination, empathy, and morality—and it acts as the repository of all that one feels. The effort to discover the biological basis for these complicated human experiences has given rise to a relatively new discipline called cognitive neuroscience. It has recently exploded as a field, due, in part, to decades of advances in neuroimaging technology that allow researchers to see the brain at work.

Mental Reserves Keep Brains Agile. Jane E. Brody. *New York Times* pF7 December 11, 2007

Despite deterioration of their brains, many elderly people maintain their mental agility into their 80s and beyond. Autopsy studies often reveal extensive brain abnormalities, similar to those associated with Alzheimer's disease, in individuals with normal cognitive ability. Scientists have suggested that a phenomenon dubbed cognitive reserve, which involves the brain's ability to develop and maintain additional neurons and connections between them, is responsible. Extra neurons and interconnections generated through mental and physical stimulation may compensate for damage to other parts of the brain. Studies have shown that education, intellectually demanding occupations, brain-stimulating hobbies, social interaction, and physical fitness can all contribute to building up cognitive reserve.

Scientists Identify the Brain's Activity Hub. Benedict Carey. *New York Times* pF6 July 1, 2008

In an article published online in *PLoS Biology*, Sporns and colleagues provide the most complete picture to date of connections within the brain's outer layer. The cerebral cortex is the brain region associated with reasoning, planning, and self-awareness. Previous MRI studies have been able to identify peaks and valleys of neural activity but indicated little about the neural networks involved. Sporns and colleagues used a new method known as diffusion spectrum imaging, which images the brain's white matter, to estimate the density and orientation of connections; they ranked the busiest locations in the cortex by number of connections and plotted them on brain maps. The results show a central focus of activity, located below the crown of the head, that is connected to more specialized brain regions in a network resembling a subway map. Although not definitive, the findings can be refined to produce a complete picture of the brain's connections.

Teens and Decision Making: What Brain Science Reveals. *New York Times Upfront* v. 140 pp18-20 April 14, 2008

Recent scientific discoveries help to explain why teens are more likely than adults to act without thinking. Until recently, scientists believed that the human brain was fully mature long before the teen years. Recent research has revealed that, although the brain reaches its full size between ages 12 and 14 years, cerebral development still has a long way to go at that time. Areas of the brain are still developing until a person reaches his early 20s. Consequently, the ways in which information is integrated by the decision making circuits in teenagers' brains are undeveloped. This could make teenagers more prone to making decisions that they may subsequently regret. Advice for teenagers on making decisions is provided, and a sidebar provides definitions of words relating to the brain.

Women and the Negativity Receptor. Aimee Lee Ball. *O: the Oprah Magazine* v. 9 pp206-9, 220 August 2008

Women's self-critical tendencies and willingness to accept negative judgments are actually caused by a part of the brain known as the anterior cingulate cortex. This part of the brain controls the observation of other people's emotions and is larger in women than in men. Hormonal shifts within the female brain make women more prone to emotional nuance and the manner in which feedback from other people is interpreted can depend upon the stage a woman is at in her menstrual cycle. Scientists agree that at least 50 percent of personality comes from the genetic materials, with the rest shaped by experiences in life. When an idea about the self is acquired environmentally, it can have influence over brain circuitry and become built into self-image. Naturally, the peer group has a great deal of sway over the way a person perceives herself. Cognitive therapy seems to address these issues by teaching the individual how to view herself in an alternate manner.

Bitmapping the Brain. Emily Masamitsu. *Popular Mechanics* v. 184 p68 November 2007

The writer provides a guide to the expanded intricate cartography of the brain that recent functional magnetic resonance imaging has allowed scientists to draw up, covering such topics as distinguishing between truth and lies, the mental and neural processes that drive economic decisions, and the brain activity linked to love.

Marriage of the Minds. Thomas Crook. *Prevention* v. 60 pp133-4 April 2008

Using functional magnetic resonance imaging, scientists have been able to observe the effects of love on the brain. When people in the early stages of infatuation are shown photos of their partners and asked to think about them, areas of the brain that are filled with the chemical dopamine are activated. Dopamine produces extremely powerful pleasurable sensations. Cocaine and amphetamine, for example, produce their effects by triggering the release of dopamine. As relationships continue, however, those areas are less responsive to the mere sight of one's partner. To be successful, the relationship must evolve from dopamine-driven euphoria to a more thoughtful cultivation of love and respect. The writer discusses ways that couples can strengthen their marriages.

Seeing into the Brain. Neena Samuel. *Reader's Digest* v. 172 p142 March 2008

Part of a special section on remarkable medical breakthroughs. The writer describes three scientific discoveries relating to the workings of the brain. U.S. scientists have, for the first time, observed the process of neurogenesis, in which a living human brain grows new brain cells. German scientists have devised a new way of using fluorescent molecules and lasers to see the whole neural network of a mouse brain in 3-D for the first time and reconstruct it on a computer, without need of a scalpel. With the help of the planet's most powerful resonance imaging machine, the 9.4 Tesla, physicians may soon be able to discover in days, rather than weeks, how well a cancer treatment is working.

Read My Slips: Speech Errors Show How Language Is Processed. Michael Erard. *Science* v. 317 pp1674-6 September 21, 2007

A number of researchers are analyzing common slips of the tongue to help understand how humans and other primates can comprehend and use language. Despite much research, scientists' understanding of how humans hear, comprehend, and speak words and sentences remains quite limited. Slips of the tongue, inadvertent linguistic errors by speakers who know the correct form, can offer clues about how the brain processes language, however, and new methods and theories are emerging from research with children, users of sign language, and even animals. Scientific interest in slips of the tongue was first sparked in the 1960s and received a major boost in the 1970s, when researchers began developing ways of eliciting many types of speech errors. Researchers now believe that analysis of the error profiles from normal speech will be an important part of efforts to understand how the many pieces of language systems are joined in the seamless manner experienced by humans.

Beyond Blood. Ewen Callaway. *Science News* v. 173 pp168-9 March 15, 2008

The next generation of MRI scanning technology could give researchers the ability to produce more detailed maps of the brain, elucidate the connections between distant regions of the brain, and diagnose diseases such as Alzheimer's disease. Almost all current functional MRI scans rely on blood flow, but many scientists argue that, although useful, blood flow can be an indirect and even misleading gauge of brain activity, particularly in view of the time lag between electrical activity and blood flow. In 2006, researchers in France and Japan claimed to have used a technique called diffusion MRI to detect the swelling of neurons as they become active; however, these results have not been repeated. Less controversially, diffusion MRI has been used to map physical connections between different brain regions. Some researchers argue that MRI could be used to detect the magnetic fields generated by the brain's electrical activity, but this would require major technological leaps.

Cells Regulate Blood Flow and Make fMRI Possible. Tina Hesman Saey. *Science News* v. 174 pp5-6 August 2, 2008

Recent research is revealing the importance of star-shaped brain cells called astrocytes. Neurons have typically received the most attention from researchers because they are the brain cells that do all the thinking, and astrocytes have been largely ignored because they were once thought to do little more than hold the brain together. Although scientists have learned in recent years that astrocytes have a hand in guiding connections between neurons and controlling levels of chemical messengers in the brain, these activities have been seen mainly as secondary. Now, Sur and colleagues and Murthy and colleagues report in the June 20 *Science* and the June 26 *Neuron*, respectively, that astrocytes regulate blood flow to parts of the brain that are thinking hard and may even play a role in storing information. Moreover, astrocytes make functional magnetic resonance imaging—which relies on the premise that blood flow is coordinated with neuron activity—possible.

The Truth About Brain Science. Robert Epstein. *Skeptical Inquirer* v. 32 pp32-3 September/October 2008

More precise scanning technologies are making it possible for scientists to monitor brain activity in real time while people are actually behaving and thinking, but although such breakthroughs are exciting, there is reason to be concerned about the hype. Claims are being made about brain research that are false, and they are being accepted without question by the press, the public, politicians, and even the courts. Most brain studies being conducted at present are correlational, which means they merely identify a link between behavior and what the brain is doing when a person is behaving. Many of these research projects do not even consider the correlation in real time and instead make assumptions that may or may not be true and go from there.

The Brain on Justice. Massimo Pigliucci. *Skeptical Inquirer* v. 32 p26, 49 September/October 2008

The fields of science, philosophy, and economics are converging to reassess how humans think and behave as social beings. The most recent development in this interdisciplinary field of inquiry is a study by Ming Hsu, Cedric Anen, and Steven Quartz of the University of Illinois and the California Institute of Technology, published in the May 23 issue of *Science*, which focuses on the areas of the brain used in the logically distinct tasks of evaluating fairness and efficiency in the distribution of resources. Using functional magnetic resonance imaging, the scientists discovered that human beings use elements of their rational brain to assess issues of optimality, whereas fairness is evaluated by emotional circuitry. Such research means that philosophers who continue their debates without considering the discoveries of science risk being falling behind in the quest to understand the character of human moral decision making.

Mental Gymnastics. Rebecca Johnson. *Vogue* v. 197 pp394-5, 425-6 December 2007

A recently emerged branch of neuroscience suggests that mental decline with age need not be a given. Scientists long believed that the brain was hard wired in its infancy and that, as people age, windows of learning opportunity seem to close, but animal studies in the early 1980s apparently indicated the presence of new neurons in the adult brain. It was not, however, until scientists from the Salk Institute could autopsy the brains of certain cancer patients who were treated with a substance that tagged all new cellular growth that they had real proof. The writer describes her quest to revitalize her brainpower, which involved having a brain scan with Daniel Amen, who believes that scanning can identify commonplace maladies, and consulting Michael Merzenich, the chief scientific officer of Posit Science, the company currently at the forefront of software designed to combat neurological deterioration.

'Neurobics' and Other Brain Boosters. Melinda Beck. *Wall Street Journal* pD1 June 3, 2008

Scientists now know that people generate new brain cells, and new connections between them, throughout life. Experts believe that the more mental reserves people build up, the better they can stave off age-related cognitive decline. A mini-industry of brain teasers, puzzles and computer games now helps worried baby boomers to challenge their brains and develop more new nerve pathways. The late neurobiologist Lawrence Katz popularized the term "neurobics" for engaging different parts of the brain to do familiar tasks. Brushing your teeth or dialing the phone with your non-dominant hand can theoretically strengthen pathways in the opposite side of the brain. Other aspects of promoting brain health are briefly discussed.

A New Approach to Treating Intractable Cases of Depression. Melinda Beck. *Wall Street Journal* pD1 October 21

A new approach to treating intractable cases of depression, known as TMS, for transcranial magnetic stimulation, is part of a new era in understanding and treating psychiatric disorders using high-tech imaging. Able to now see depression in the brain, scientists stimulate the left prefrontal cortex, the area linked to depression. TMS treatment uses rapid magnetic pulses to send a mild electrical current to the area for about 40 minutes, every day for four to six weeks. If one suffers from major depression, relief could come within a few weeks, with improvement in mood, sleep, appetite, energy level and restoration of hopefulness and self-esteem.

Index